Introductory Eigenphysics

Introductory Eigenphysics

An Approach to the Theory of Fields

C. A. Croxton

*Cavendish Laboratory, Department of Physics,
and Fellow, Jesus College, Cambridge*

JOHN WILEY & SONS

London · New York · Sydney · Toronto

Library of Congress Cataloging in Publication Data:

Croxton, C. A.
Introductory Eigenphysics.

1. Field theory (Physics)   2. Boundary value problems.
I. Title.

QC174.45.C76   530.1′4   74–175
ISBN 0 471 18929 4 (Cloth bound)
ISBN 0 471 18930 8 (Paper bound)

Printed in Great Britain by J. W. Arrowsmith Ltd., Winterstoke Road, Bristol

Preface

Compartmentalization of the material appearing in present day physical sciences curricula is an inevitable consequence of several effects, not least of which is the rapid research development, and the desire to update and incorporate as many important advances as possible. This is certainly to be encouraged—undoubtedly such topical revision of the course can only stimulate enthusiasm—although it can, and often does leave no opportunity to stop off and relate to other aspects of the subject. This book is concerned with boundary value phenomena—not least of all with boundaries between compartments, for in the author's experience all too many undergraduates fail to recognize the common features of superficially distinct problems in physics. Indeed, one of the particularly satisfying features of the subject is the explanation of a large number of phenomena with a minimum of physical or mathematical apparatus. What is not to be encouraged is the all too familiar undergraduate standpoint that all these problems are distinct, and to be approached *ab initio*. So, this book is meant to be something other than a catalogue of mathematics for physicists, even less a catalogue of physics for mathematicians. What we attempt here is to break down the conventional divisions between quantum and classical mechanics, between physical optics and electron scattering, between electrostatics and aerodynamics, and so on, and, as we do so, develop a confidence in the application of the techniques discussed in this book. The principal boundary is often to be found in the almost complete failure of the undergraduate to relate or apply the equipment developed in his mathematics course to his physics. For example, a student will often show the greatest reluctance to plunge into a surface spherical harmonic analysis of an angular distribution, even though he might have 'done it in maths'. However, suggest to him that it need be no more disconcerting than a routine Fourier analysis of a periodic function, and his bearings are fixed. This is, perhaps, a criticism which can be levelled at the several mathematical physics texts: to an undergraduate physics student their approach can appear sterile, and can suffer from a surfeit of rigour at the expense of physical understanding and application.

In this book the mathematics is regarded quite firmly as the tool of physics. Again, all too often, a student encountering the wavemechanical formulation of the hydrogen atom for the first time recalls only that he was simultaneously confronted with the solution of the Legendre and Laguerre differential equations. Admittedly the various quantum numbers arise quite naturally in the course of the analysis, and the apparently *ad hoc* formulation of Bohr is, of course, placed on a sound analytic basis. The quantization, and indeed much of the physics, can however be obtained without actually *solving* the various differential equations which arise: indeed, in this case we need only require the wavefunction to be continuous, and to vanish at infinity. In consequence the solution to these and the several other differential equations which feature throughout physics are firmly relegated to appendices at the back of the book. Of course, some relevant mathematical elaboration may be appropriate, and in this case the digression will also appear in the index.

It is no coincidence that so few differential equations describe so much physics, and it is a pity to neglect and not fully exploit this unifying basis to the subject. This field approach to the solution of physical problems may then be reduced to a more or less mechanical task, and the distinguishing features appear in the nature and values of the boundary conditions, and in the symmetry and degeneracy of the system.

The range of physical problems discussed is deliberately varied, and there is no logical development other than through the symmetry of the coordinate systems. This is not so much due to the irascibility of the author as to emphasize the ubiquity and unity of the field approach. Of course, only a very limited number of problems succumb to exact solution, even if a separation of variables is effected. A chapter on the principal methods of approximate solution is therefore included, and application to various problems in geophysics and quantum chemistry and mechanics are discussed. Greater emphasis could perhaps have been placed on these approximate methods, especially since virtually no physical system of interest is susceptible to exact solution. On the other hand, the objective of this text is quite clearly to present a physical approach to some of the principal mathematical functions, and at undergraduate level the emphasis must surely be on the exact solutions—only then can the extension to orthonormal series descriptions to perturbed situations be considered.

Chapters 2, 3, 4 and 5 are concluded with a number of problems which are designed to both supplement and complement the text. These examples are primarily of degree level and have largely been taken from recent first- and second-year Tripos examination papers.

Obviously in the production of such a book as this one has inevitably benefited from many of the standard texts, and the subjects treated here, except for the development of a small number of results, are to be found in the published literature, although some revision and infill is necessary to make possible a presentation at an elementary level. Particularly useful sources have been listed in a bibliography, although the list is by no means exhaustive.

The majority of this book was written whilst on leave at the Centre for Advanced Studies in Physics, Delhi University and at the Raman Research Institute, Bangalore, India: the author acknowledges with pleasure the social and environmental facilities at these institutions which have helped make this book possible. Much of the content of this book has been developed in the course of undergraduate teaching, when students have assiduously pointed out difficulties, inconsistencies and vagaries in the presentation and development, the isolation and discussion of which will, hopefully, enhance the value of the book for those for whom it is primarily intended. It is a pleasure to record my thanks to all those who contributed in this way. Errors and omissions doubtless remain, and suggestions from users of the book will be particularly appreciated. I am also indebted to Professor Brian Pippard and to the Cambridge University Press for permission to extract a number of the examples from *Cavendish Problems in Classical Physics* and from recent Part IA and IB Natural Sciences Tripos Examination papers. It is a particular pleasure to acknowledge the generous contributions of Owen Saxton and Martyn Horner in the early days of the manuscript: to them I extend my warmest thanks.

If the reader finds in the logical texture of this book a simplifying and uni-fying basis, the author will have achieved his aim.

Cavendish Laboratory,   Clive A. Croxton
Cambridge,
November 1973

Contents

CHAPTER 1

The Field Equations

1.1 Introduction

In classical electrostatics and gravitation we may always express the spatial variation of the relevant potential function $\psi(\mathbf{r})$ in differential form: the potential field may then be determined by a variety of mathematical techniques depending upon the nature of the equation, and the system of coordinates. The final solution to the differential equation yields the spatial distribution of the electrostatic or gravitational potential, this subject of course to certain boundary conditions. The equation principally involved in electrostatic and gravitational problems is the Poisson equation, $\nabla^2 \psi(\mathbf{r}) = f(\mathbf{r})$, and its homogeneous variant $\nabla^2 \psi(\mathbf{r}) = 0$, the Laplace equation. $f(\mathbf{r})$ may be understood as the spatial distribution of an electrostatic or gravitational 'source' or 'sink' of charge or matter within the region over which the Poisson equation is satisfied. It is principally in terms of gravitation and electrostatics that we have come to understand the terms potential and field. If, however, we are prepared to extend our definition and consider *generalized potentials* in the more abstract sense together with their associated *generalized flux* or *force field*, then we shall find that very similar equations arise in many branches of physics, not just electrostatics and gravitation. Moreover, the solutions to these equations yielding the spatial distribution of the generalized potential will be functionally very similar, and will differ only in the nature of the boundary conditions and the physical interpretation placed upon the generalized parameters of the solution. Thus, our ability to solve a given field equation enables us to simultaneously deal with a whole range of physical phenomena, apparently unrelated, but which are nevertheless to be analytically associated by the form of the field equation. By way of example we might cite the distribution of amplitude $\psi(r)$ of standing surface waves on a circular trough of water. The distribution of amplitude about the mean is described by Bessel's equation, as is the distribution of alternating current $\mathbf{j}(r, t)$ over the cross-section of a cylindrical metallic conductor. We could extend this particular example to include the radial distribution of amplitude $A(r)$ in the Fraunhofer diffraction of monochromatic light by a circular pinhole, and the normal modes of vibration of a circular drumhead. The generalized potential, the amplitude, may in each case be described in terms of the Bessel function, but the physical interpretation of the potential function differs in each situation. What we should also notice about these examples is that the symmetry of the problem is in every case cylindrical. We shall find that the symmetry and dimensionality of the problem will invariably play a large part in determining the functional form of the potential field which develops, and this will provide a useful way of categorizing the kinds of solution which arise in later chapters.

It is the purpose of this book in general, and the next few sections in particular, to establish and solve the more important types of field equation generally encountered in physics. We shall indicate how these field equations have been solved in the various coordinate systems, and shall draw widely from various branches of physics, the unifying aspect being the method of solution rather than the physical nature of the problem.

1.2 Poisson and Laplace's Equation

A particularly important class of fields $\mathbf{I}$ are characterized by their being *conservative*. The work done in moving a particle along an infinitesimal element of displacement dl in the continuous field $\mathbf{I(r)}$ depends only upon the initial and final positions of the particle, and is wholly independent of the sequence or nature of the intervening displacements. Such a field $\mathbf{I}$ is termed *irrotational* since its circulation vanishes

$$\oint \mathbf{I} \cdot d\mathbf{l} = 0 \qquad (1.2.1)$$

that is, the work done in taking the particle along a closed loop in the field is zero. Provided the irrotational field $\mathbf{I(r)}$ and its first derivatives are continuous then from Stokes' theorem we may relate the line and surface integrals thus:

$$\oint \mathbf{I} \cdot d\mathbf{l} = \oint \text{curl}\, \mathbf{I} \cdot d\mathbf{A} = 0 \qquad (1.2.2)$$

where $\mathbf{A}$ in the second integral is the area enclosed by the contour in the first. We therefore conclude that for an irrotational field, curl $\mathbf{I} = 0$. Obvious examples of fields in class $\mathbf{I}$ (irrotational) include the gravitational and electrostatic fields, although generally some dissipative agent such as friction will intervene and the field will contain a non-irrotational component. $\mathbf{I}$ will, of course, remain irrotational, and for this component alone the circulation will vanish. In practice, however, it is difficult to separate the two components, and the closed line integral will generally be > 0. The magnetostatic field is an example of a *non*-irrotational field in the presence of currents, and we have

$$\oint \mathbf{H} \cdot d\mathbf{l} = i \quad \text{(Ampère's circuital theorem)}$$

where i is the current enclosed by the contour of the line integral. Otherwise, of course, if the contour encloses no currents, the magnetic field is irrotational.

Since the line integral of $\mathbf{I} \cdot d\mathbf{l}$ for an irrotational field depends only on the initial and final points, $\mathbf{I(r)}$ may be related to a scalar potential distribution $\psi(\mathbf{r})$, as follows. The potential energy $\psi(\mathbf{r})$ is the work done in moving the particle in an irrotational field from the point $\mathbf{r}$ to some datum point $\mathbf{r}_0$:

$$\psi(\mathbf{r}) = \int_{\mathbf{r}}^{\mathbf{r}_0} \mathbf{I} \cdot d\mathbf{l} + c \qquad (1.2.3)$$

where c is a constant scalar.

The change in potential energy in moving from $\mathbf{r}_1$ to $\mathbf{r}_2$ is therefore

$$\psi(\mathbf{r}_1) - \psi(\mathbf{r}_2) = \int_{\mathbf{r}_1}^{\mathbf{r}_0} \mathbf{I} \cdot d\mathbf{l} - \int_{\mathbf{r}_2}^{\mathbf{r}_0} \mathbf{I} \cdot d\mathbf{l} \qquad (1.2.4)$$

This is, the work done in a conservative system is equal to the decrease in the potential energy. If now the two points (x, y, z), $(x + \delta x, y, z)$ are infinitesimally separated, we may write

$$\delta\psi = -\int_{x}^{x+\delta x} I_x \, dx$$

i.e.

$$I_x = -\frac{\partial\psi}{\partial x}$$

Thus, if the irrotational field $\mathbf{I}(\mathbf{r})$ is defined by equation (1.2.1) we may relate it to the gradient of a distribution of scalar potential:

$$\mathbf{I}(\mathbf{r}) = -\nabla\psi(\mathbf{r}) \qquad (1.2.5)$$

We might, therefore, have anticipated that for irrotational fields curl $\mathbf{I} = 0$, since we have the vector identity for a scalar field $\psi : \nabla \wedge (\nabla\psi) = 0$. It follows that on an equipotential surface, $\psi(\mathbf{r}) = $ constant, the field vector $\mathbf{I}(\mathbf{r})$ will be everywhere normal, there being no field component parallel to the equipotential surface.

A second important class of fields $\mathbf{S}(\mathbf{r})$ are characterized by their being *solenoidal*. We shall find that in this case the field $\mathbf{S}(\mathbf{r})$ may be related to a *vector potential* $\boldsymbol{\psi}(\mathbf{r})$ in an analogous way to the scalar potential relationship in equation (1.2.5). Solenoidal fields are characterized by the zero divergence of the vector field $\mathbf{S}$, or alternatively, that there is zero net flux density crossing a closed surface surrounding the point $\mathbf{r}$, thus:

$$\oint \mathbf{S}(\mathbf{r}) \cdot \hat{\mathbf{n}} \, dA = \int_V \operatorname{div} \mathbf{S}(\mathbf{r}) \, dV = 0 \qquad (1.2.6)$$

where $\hat{\mathbf{n}}$ is the unit normal vector.

This is, of course, Gauss' divergence theorem, and the solenoidal condition, $\operatorname{div} \mathbf{S}(\mathbf{r}) = 0$, tells us that there is no net source–sink distribution within the volume element centred at $\mathbf{r}$. Since the divergence of the curl of any vector field is identically zero, it follows that we may represent a solenoidal field as

$$\mathbf{S}(\mathbf{r}) = \nabla \wedge \boldsymbol{\psi}(\mathbf{r}) \qquad (1.2.7)$$

where $\psi(\mathbf{r})$ is termed the *vector potential*. Of course, the vector field is not uniquely specified in this way, for we could supplement the vector potential ψ by the gradient of a scalar potential. Then we should have

$$\mathbf{S}(\mathbf{r}) = \nabla \wedge (\psi(\mathbf{r}) + \nabla\psi(\mathbf{r})) \tag{1.2.8}$$

and since $\nabla \wedge (\nabla\psi) = 0$ the field $\mathbf{S}$ remains unchanged. This is known as *gauge invariance* and the transformation $\psi \rightarrow \psi + \nabla\psi$ is known as a *gauge transformation*. A comparison of the scalar and vector potentials is given in Table 1.2.1.

Table 1.2.1. Comparison of scalar and vector potentials

Rotational fields	Solenoidal fields
$\oint \mathbf{I} \cdot \mathbf{dl} = 0$	$\int \nabla \cdot \mathbf{S} \, d\mathbf{V} = 0$
$\mathbf{I} = -\nabla\psi$	$\mathbf{S} = \nabla \wedge \psi$
$\nabla \wedge \mathbf{I} = 0$	$\nabla \cdot \mathbf{S} = 0$

We shall now derive the Laplace and Poisson differential equations for scalar and vector potential distributions. These equations, as we shall see, are of central physical importance, and will preoccupy us throughout this book.

To be quite general, we shall consider the field $\mathbf{F}(\mathbf{r}) = \mathbf{I}(\mathbf{r}) + \mathbf{S}(\mathbf{r})$ containing both irrotational and solenoidal components. All we otherwise assume is that the fields are single valued, continuous and everywhere differentiable. Taking the divergence of the field $\mathbf{F}(\mathbf{r})$, we have

$$\nabla \cdot \mathbf{F}(\mathbf{r}) = \nabla \cdot [\mathbf{I}(r) + \mathbf{S}(r)] = \nabla \cdot \mathbf{I}(r) \tag{1.2.9}$$

since $\nabla \cdot \mathbf{S}(\mathbf{r})$ is, by definition, zero. From equation (1.2.5) we have

$$\nabla \cdot \mathbf{F}(\mathbf{r}) = -\nabla \cdot (\nabla\psi(\mathbf{r})) = -\nabla^2\psi(\mathbf{r})$$

or

$$\nabla^2\psi(\mathbf{r}) = f(\mathbf{r}) \tag{1.2.10}$$

which is Poisson's equation for a scalar potential distribution. $f(\mathbf{r})$ represents the spatial distribution of the divergence of the vector field $\mathbf{F}(\mathbf{r})$, and this is generally associated with the scalar source density. When the source density is zero, that is, when the divergence of the vector field is zero, then we obtain the homogeneous form of equation (1.2.10), termed Laplace's equation.

The scalar function $f(\mathbf{r})$ is an intrinsic characteristic of the field $\mathbf{F}(\mathbf{r})$ at any point, and may be related to the *curvature* of the potential surface $\psi(\mathbf{r})$. In the case of Laplace's equation the absence of any net distribution of sinks and sources ensures that $\mathbf{F}(\mathbf{r})$ has zero divergence, and that correspondingly the surface $\psi(\mathbf{r})$ possesses no curvature, although it will generally possess a gradient. A potential surface satisfying the Poisson equation, however, will develop a curvature directly related to the scalar distribution $f(\mathbf{r})$, or the divergence of the field $\mathbf{F}(\mathbf{r})$. In the electrostatic case we may consider lines of flux originating from positive point charges (sources), and terminating on negative charges (sinks), whereupon we obtain a potential distribution of the form shown in Figure 1.2.1 for charge distributions $f_+(\mathbf{r}), f_-(\mathbf{r})$.

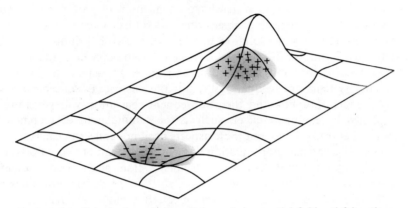

Figure 1.2.1 Schematic variation of the scalar potential field satisfying the Poisson equation for charge distributions $f_+(\mathbf{r}), f_-(\mathbf{r})$

The uniqueness with which we are able to specify the potential distribution from the Poisson and Laplace equations will be discussed in Section 1.5.

If we take the curl of the field $\mathbf{F}(\mathbf{r})$, we have

$$\nabla \wedge \mathbf{F}(\mathbf{r}) = \nabla \wedge [\mathbf{I}(\mathbf{r}) + \mathbf{S}(\mathbf{r})] = \nabla \wedge \mathbf{S}(\mathbf{r}) \qquad (1.2.11)$$

since the curl of an irrotational field is zero. From equation (1.2.7) we have

$$\nabla \wedge \mathbf{F}(\mathbf{r}) = \nabla \wedge (\nabla \wedge \psi(\mathbf{r}))$$
$$= \nabla(\nabla \cdot \psi(\mathbf{r})) - \nabla^2 \psi(\mathbf{r}) \qquad (1.2.12)$$

from the vector identity. We see that to uniquely specify the field $\mathbf{F}(\mathbf{r})$ we have to specify the divergence of the vector potential $\psi(\mathbf{r})$. Since from equation (1.2.11) we see that the field is solenoidal, it is reasonable to choose $\nabla \cdot \psi = 0$, whereupon equation (1.2.12) becomes

$$\nabla^2 \psi(\mathbf{r}) = \mathbf{f}(\mathbf{r}) \qquad (1.2.13)$$

which is Poisson's equation for a vector potential distribution. This particular choice, $\nabla \cdot \psi = 0$, is termed the *Coulomb gauge*. $\mathbf{f}(\mathbf{r})$ represents the spatial distribution of the curl of the vector field $\mathbf{F}(\mathbf{r})$, and this is generally associated with the vector source density. We regain the Laplace equation if curl $\mathbf{F}(\mathbf{r}) = \mathbf{f}(\mathbf{r})$ is zero:

$$\nabla^2 \psi(\mathbf{r}) = 0 \qquad (1.2.14)$$

1.3 The Diffusion Equation

The diffusion equation may be extensively applied in many branches of physics, and generally arises when there is any spatial gradient of concentration, temperature, etc. The equation applies in the macroscopic limit, and to this extent the conduction of heat in a solid may be accounted for in terms of a continuous diffusion process, although we know that at a molecular level the distribution of thermal energy proceeds as a phonon–phonon interaction process within the lattice. The evolution of a density concentration of particles in a dense fluid is an extremely complex problem which as yet has no entirely satisfactory solution, and certainly no solution in terms of a classical diffusion process. Nevertheless, asymptotically in time, and in the macroscopic limit, the process is well-described by the diffusion equation and therefore assumes an important place in our system of field equations. Again, the existence of concentrations of density, temperature, etc. and their subsequent relaxation will generally be accompanied by conjugate effects such as the simultaneous development of thermal and entropy fluxes, but these aspects need not concern us here.

It is clear that the diffusion equation will differ from that of either Poisson or Laplace in that it incorporates a time dependence. If the generalized gradient of concentration is sustained by some means so that the distribution does not relax, then we shall find that we recover an equation of Laplacian form.

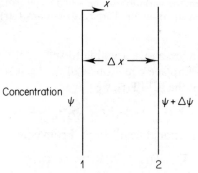

Figure 1.3.1 Geometry for the calculation of a generalized diffusive flux

We may develop the diffusion equation by considering the flux of particles from surface 1 to surface 2 arising from the concentration gradient in the medium. Clearly the net flux of particles in the x direction is given as

$$\mathscr{F}(x) = -D\frac{\Delta\psi}{\Delta x} \tag{1.3.1}$$

and this is developed between surfaces 1 and 2 separated by the spatial increment Δx. D is a constant coefficient of proportionality. Provided the incremental distance Δx is small we may Taylor expand the concentration variation and write to a first approximation

$$\Delta\psi \cong +\frac{\partial\psi}{\partial x}\Delta x, \tag{1.3.2}$$

so that in conjunction with equation (1.3.1) we have, assuming the region is isotropic

$$\mathscr{F} = -D\nabla\psi \tag{1.3.3}$$

where D is now identified as the diffusion coefficient. Equation (1.3.3) is a statement of Fick's Law, i.e. that the particle flux is directly proportional to the concentration gradient. As we have observed, significant deviations from this equation arise at a molecular level especially in the initial stages of the evolution, but even at this level equation (1.3.3) is asymptotically correct, and is certainly satisfactory at the macroscopic level over small particle concentrations and variations in density and temperature. Fick's Law relates the flux of diffusing particles to the generalized potential gradient $\nabla\psi$; to establish the time-dependence of the diffusion equation we utilize the conservation condition according to which the divergence of the diffusing flux at a point is

$$\nabla \cdot \mathscr{F} = -\frac{\partial \psi}{\partial t} + c - a \tag{1.3.4}$$

The terms c and a represent creation (source) and annihilation (sink) terms in the potential field ψ. Equation (1.3.4) is then self-evident. Taking the divergence of Fick's equation (1.3.3) and combining the result with equation (1.3.4) we immediately obtain the general diffusion equation:

$$\nabla^2\psi(\mathbf{r}, t) = \frac{1}{D}\frac{\partial\psi(\mathbf{r}, t)}{\partial t} + a(\mathbf{r}, t) - c(\mathbf{r}, t) \tag{1.3.5}$$

As we anticipated, under certain circumstances ($a, c = 0$, $\partial\psi/\partial t = 0$; $|a| = |c|$, $\partial\psi/\partial t = 0$) the diffusion equation reduces to Laplace's equation. If the distribution is purely stationary ($\partial\psi/\partial t = 0$) then equation (1.3.5) may reduce to Poisson form.

We see that if we assume separability of the variables in the distribution, $\psi(\mathbf{r}, t) = R(\mathbf{r})T(t)$ where R and T are purely functions of position and time respectively, then for no net distribution of sources or sinks we have

$$DT\nabla^2 R = R\dot{T}$$

i.e.

$$\frac{\nabla^2 R}{R} = D^{-1}\frac{\dot{T}}{T} \tag{1.3.6}$$

This being so we can quite obviously set each side of equation (1.3.6) equal to a constant $-h$, since each side is a function of a single (different) variable. We then have the separated equations

$$\frac{\nabla^2 R}{R} = -h$$

$$\frac{\dot{T}}{T} = -hD \tag{1.3.7}$$

$-h$ is termed the separation constant, and may be determined from the boundary conditions. We need not have distributed D amongst the equations as we have in (1.3.7), nor need we have chosen the separation constant to be $-h$: these particular choices will, however, be convenient for later purposes. From equations (1.3.7) the relaxation in concentration would appear to be of irreversible exponential form:

$$T(t) = T_0 \exp(-hDt) \tag{1.3.8}$$

where $(hD)^{-1}$ has the significance of a characteristic relaxation time for the evolution.

1.4 The Wave Equation

Perhaps the most important of the field equations, the wave equation appears in virtually every aspect of classical and quantum physics, and its solution under various conditions will constitute a significant portion of this book. Whilst in the wave equation

$$\nabla^2 \psi(\mathbf{r}, t) = \frac{1}{c^2} \frac{\partial^2 \psi(\mathbf{r}, t)}{\partial t^2} \tag{1.4.1}$$

c has in every case the dimensions of velocity, the physical quantities in which it is expressed will vary from problem to problem, and these we shall discuss as the situation arises. c nevertheless does represent the velocity of propagation of the wave in the medium.

Again, writing $\psi(\mathbf{r}, t) = R(\mathbf{r})T(t)$ we are able to effect a separation of the variables as follows: writing equation (1.4.1) in the form

$$T(t)\nabla^2 R(\mathbf{r}) = \frac{R(\mathbf{r})}{c^2} \frac{d^2 T(t)}{dt^2} \tag{1.4.2}$$

$$\nabla^2 R + k^2 R = 0 \tag{1.4.3}$$

$$\frac{d^2 T}{dt^2} + c^2 k^2 T = 0 \tag{1.4.4}$$

where k^2 is again a separation constant, termed the *eigenvalue*, and is determined in the course of solution of the *Helmholtz Equation*, (1.4.3). k may in fact be wholly real, in which case solutions to equation (1.4.4) will be oscillatory and of the form

$$T(t) = T_0 \exp(\pm ickt) \tag{1.4.5}$$

or wholly imaginary, in which case the solutions will be simply exponential, and of the form

$$T(t) = T_0 \exp(\pm kct) \tag{1.4.6}$$

We see from these solutions that kc must have units of frequency. k has the dimension of reciprocal length, and represents the *wave number* of the oscillation:

$k = \omega/c$ where ω is the angular frequency. The wave equation (1.4.3) may however, contain a damping term as follows, corresponding to a complex separation constant

$$\nabla^2 \psi(\mathbf{r}, t) = \frac{1}{c^2} \frac{\partial^2 \psi(\mathbf{r}, t)}{\partial t^2} + \beta \frac{\partial \psi(\mathbf{r}, t)}{\partial t} \tag{1.4.7}$$

where β is a *damping* or *loss coefficient*. A separation of variables may again be effected and we obtain the separated equations

$$\nabla^2 R + k^2 R = 0 \tag{1.4.8a}$$

$$\frac{d^2 T}{dt^2} - c^2 \beta \frac{dT}{dt} + c^2 k^2 T = 0 \tag{1.4.8b}$$

We observe first of all that the Helmholtz equation (1.4.8a) is of the same form as its undamped counterpart equation (1.4.3). The time dependence is given as the solution of equation (1.4.8b):

$$T(t) = T_0 \exp(-\beta c^2 t/2) \exp[\pm \sqrt{(\beta^2 c^4/4 - k^2 c^2)}t] \tag{1.4.9}$$

which represents a damped oscillation in time. We notice that if $\beta = 0$ (no damping) we recover the solutions (1.4.5), (1.4.6) according as k is wholly real or wholly imaginary. For values of the damping constant such that $c^2 \beta^4/4 > k^2$ (cf. equation (1.4.9)) then the solution (1.4.9) again becomes purely exponential and for decaying solutions is termed the *overdamped* case. This clearly corresponds to the physical situation wherein significant exponential damping has occurred *within* the period of the first oscillation. No oscillation occurs therefore, and this may be confirmed by taking the first derivative of the time solution (1.4.9) subject to the condition $\beta^2 c^2/4 > k^2$; we find that there are no stationary points ($dT/dt = 0$).

In most physical situations the 'explosive' solutions arising from the exponential increase in time of the solution $T(t)$ are rejected. We generally require well-behaved solutions to our differential equations, that is, solutions which are finite, single-valued and everywhere differentiable. However, of the infinity of possible solutions which satisfy the various field equations developed above, most have to be rejected when we come to consider the boundary conditions.

1.5 Boundary Conditions and Uniqueness of Solution

Any one of the second-order differential field equations discussed in the previous sections will, of course, have an infinity of possible solutions. It may be that the precise functional form of the solution can be partially established on physical grounds, or on the grounds that the solution should be well-behaved and not develop infinities, singularities or discontinuities. The final selection of the relevant solution is effected by the application of *boundary conditions*. Thus, if we have information concerning the value or derivative of the unknown function on a closed boundary of the region then we may often determine the arbitrary constants appearing in the general solutions of the Laplace and Helmholtz equations. In the case where the value of the function is specified over the boundary we have the *Dirichlet problem*. If on the other hand the gradient is specified, then the boundary condition is termed the *Neumann problem*. We may also have mixed (*Cauchy*) boundary conditions in which both the value and the gradient are specified at the boundary. Laplace's equation, for example, arises frequently in hydrodynamic, gravitational and electrostatics problems. In the latter case the surfaces of conductors naturally constitute the boundaries of the problem, and their potentials represent Dirichlet conditions. In problems of steady irrotational hydrodynamic flow of an incompressible fluid the velocity potential must satisfy Neumann conditions at the boundary in accordance with the simple physical requirement that the fluid cannot flow into the solid boundary.

We may conveniently consider the uniqueness of the solutions to the field equations using Green's identity for two well-behaved functions ϕ_1 and ϕ_2:

$$\int_\tau (\nabla\phi_1 \cdot \nabla\phi_2 + \phi_1 \nabla^2\phi_2)\,d\tau = \oint_A \phi_1(\nabla\phi_2)_n\,dA \tag{1.5.1}$$

where the first integral is over the volume τ and the second is over the area A enclosing the volume τ. $(\nabla\phi_2)_n$ represents the normal gradient at the surface A. Suppose now we consider that we have two solutions to Laplace's equation ψ_1 and ψ_2 in the region τ, bounded by A, so that $\nabla^2\psi_1 = 0, \nabla^2\psi_2 = 0$. Moreover, we shall suppose that these solutions satisfy Dirichlet boundary conditions on the surface A. In that case $\psi_1 - \psi_2 = 0$ at A, since both ψ_1 and ψ_2 must satisfy the boundary condition. If we substitute the function $(\psi_1 - \psi_2) = \phi_1 = \phi_2$ in equation (1.5.1) we obtain

$$\int_\tau \{[\nabla(\psi_1 - \psi_2)]^2 + (\psi_1 - \psi_2)(\nabla^2\psi_1 - \nabla^2\psi_2)\}\,d\tau$$
$$= \oint_A (\psi_1 - \psi_2)\nabla(\psi_1 - \psi_2)_n\,dA \tag{1.5.2}$$

The right-hand side of equation (1.5.2) is zero since the integral is evaluated over the boundary surface A. The left-hand side reduces to

$$\int_\tau [(\psi_1 - \psi_2)]^2\,d\tau = 0 \tag{1.5.3}$$

whereupon we must have, since the integrand is everywhere positive, $\nabla(\psi_1 - \psi_2) = 0$, i.e.

$$\psi_1 - \psi_2 = \text{constant} \tag{1.5.4}$$

Now we have already observed that $\psi_1 - \psi_2 = 0$ on the boundary (Dirichlet condition) and so it follows from equation (1.5.4) that $\psi_1 = \psi_2$. In other words the solution is uniquely defined. If Neumann boundary conditions applied in the solution of Laplace's equation so that $\nabla\psi_1 = \nabla\psi_2$ on the boundary, then the right-hand side of equation (1.5.2) again vanishes and we again obtain

$$\psi_1 - \psi_2 = \text{constant}$$

throughout the volume. However since at the boundary $\nabla(\psi_1 - \psi_2) = 0$, we are left with an undetermined constant. In this way it may be shown that in general the solution to Poisson's equation and the Helmholtz equation are *not* unique.

As far as application of the uniqueness theorem to Laplacian potential distributions $\psi(\mathbf{r})$ is concerned, however determined, *any* potential which

   (i) has the assigned values along certain boundaries
   (ii) has the assigned discontinuities (which may be zero) in $(\nabla\psi)_n$ at the boundaries
   (iii) vanishes 'sufficiently rapidly' (if required) at infinity

represents *the* potential distribution.

1.6 Orthogonality and the Sturm–Liouville Equation

We shall find that all the differential equations developed in this book have the general form

$$\frac{d^2\psi_n(x)}{dx^2} + f_1(x)\frac{d\psi_n(x)}{dx} + [f_2(x) + k_n^2 f_3(x)]\psi_n(x) = 0 \qquad (1.6.1)$$

where ψ_n is an eigenfunction satisfying the differential equation, k_n is the appropriate separation constant or eigenvalue and $f_1(x)$, $f_2(x)$ and $f_3(x)$ are arbitrary functions which vary from problem to problem. Equation (1.6.1) may be rewritten in perfect differential form: multiplying throughout by $u(x) = \exp(\int f_1(x)\,dx)$, equation (1.6.1) becomes

$$\frac{d}{dx}\left\{u(x)\frac{d\psi_n}{dx}\right\} + (h_1(x) + k_n^2 h_2(x))\psi_n = 0 \qquad (1.6.2)$$

where $h_1(x) = u(x)f_2(x)$, $h_2(x) = u(x)f_3(x)$. All the field equations which we shall encounter may be reduced to Sturm–Liouville form, and the following development is therefore of general application throughout this book.

We may equally consider a second eigenfunction $\psi_m(x)$ having the corresponding eigenvalue k_m:

$$\frac{d}{dx}\left\{u(x)\frac{d\psi_m}{dx}\right\} + (h_1(x) + k_m^2 h_2(x))\psi_m = 0 \qquad (1.6.3)$$

Multiplying equation (1.6.2) by ψ_m and equation (1.6.3) by ψ_n and subtracting, we obtain

$$\frac{d}{dx}\{u(\psi_n'\psi_m - \psi_m'\psi_n)\} + h_2(k_n^2 - k_m^2)\psi_n\psi_m = 0 \qquad (1.6.4)$$

Integrating equation (1.6.4) over the interval $a \leqslant x \leqslant b$ we obtain

$$\{u(\psi'_n\psi_m - \psi'_m\psi_n)\}_a^b = (k_m^2 - k_n^2)\int_a^b h_2(x)\psi_n(x)\psi_m(x)\,dx \qquad (1.6.5)$$

Now, if the left-hand side is zero, and $km^2 \neq kn^2$, then

$$\int_a^b h_2(x)\psi_n(x)\psi_m(x)\,dx = 0 \qquad (1.6.6)$$

In words, the functions $\psi_n(x)$, $\psi_m(x)$ are orthogonal, with respect to the weighting function $h_2(x)$, over the interval a, b. The interval is prescribed by the vanishing of $[u(\psi'_n\psi_m - \psi'_m\psi_n)]_a^b$ (equation (1.6.5)). Moreover, if

$$\int_a^b h_2(x)\psi_m(x)\psi_m(x)\,dx = N_m \qquad (1.6.7)$$

then we may expand an arbitrary function $g(x)$ as a 'Fourier' series in terms of the *complete orthogonal set* of eigenfunctions, thus:

$$g(x) = \sum_{m=0}^{\infty} A_m\psi_m(x) \qquad (1.6.8)$$

where the coefficients A_m may be determined directly from equation (1.6.8) as

$$\frac{1}{N_m}\int_a^b g(x)h_2(x)\psi_m(x)\,dx = A_m \qquad (1.6.9)$$

Alternatively we may slightly redefine our eigenfunctions such that

$$\int_a^b h_2(x)\psi_n(x)\psi_m(x)\,dx = \delta_{mn} \qquad (1.6.10)$$

where the Krönecker delta δ_{mn} is unity for $n = m$, and zero for $n \neq m$. In this case we may expand the arbitrary function in terms of the *complete orthonormal set* (the functions now being normalized) when the coefficients are given as

$$\int_a^b g(x)h_2(x)\psi_m(x)\,dx = A_m \qquad (1.6.11)$$

The significance of these comments will be quite clear when it is recalled that the *sum* of eigenfunctions satisfying the differential equation is also a solution, and in fact represents the general solution (equation (1.6.8)). We may therefore attempt to represent any arbitrary function $g(x)$ over the interval a, b in terms of a complete set of eigenfunctions where the coefficients are given by equation (1.6.9) or equation (1.6.11). Orthonormal expansions will be made throughout this book in terms of a whole variety of functions depending in general upon the symmetry of the coordinate system. Perhaps the simplest and most familiar example is the Fourier expansion of $g(x)$ in terms of the trigonometric functions:

$$g(x) = \sum_{m=0}^{\infty} (A_m \sin mx + B_m \cos mx) \qquad (1.6.12)$$

in which case the coefficients A_m and B_m are given as

$$\left.\begin{aligned} \frac{1}{\pi} \int_0^{2\pi} g(x) \sin mx \, \mathrm{d}x &= A_m \\[2mm] \frac{1}{\pi} \int_0^{2\pi} g(x) \cos mx \, \mathrm{d}x &= B_m \end{aligned}\right\} \qquad (1.6.13)$$

where $N_m = \pi$, $h_2(x) = 1{\cdot}00$, and the limits a, b are $0, 2\pi$.

It is important to emphasize that the set must be *complete*, and physically this corresponds to the requirement that *all possible* normal modes or eigenfunctions of the system must be 'accessible' or available for excitation. The class of functions and the adequacy with which they may be represented by an eigenfunction expansion is discussed in some detail by Margenau and Murphy.

When a function is expanded in terms of a complete set of orthonormal functions $\psi_n(x)$ thus,

$$f(x) = a_1\psi_1(x) + a_2\psi_2(x) + \cdots + a_n\psi_n(x) + \cdots$$

the components may be regarded as orthogonal coordinate vectors in a space of infinitely many, continuous dimensions. The orthonormal set plays the role of a set of orthogonal unit vectors where the expansion coefficients represent the vector components in an analogous way that an n-dimensional vector can be written as the sum of its components along the coordinate axes. This unified approach to orthonormal expansions in terms of the algebraic properties of vectors in *Hilbert space* differs from the finite-dimensional Euclidean space in that a vector may now have infinitely many mutually orthogonal components. The evolution of a dynamical variable $f(x, t)$ in Hilbert space is now seen as an evolution amongst the coefficients $a_n(t)$ of the orthonormal expansion.

In the most general formulation of field problems, which we do not discuss in this book, the state function $f(x, t)$ is represented by a vector in Hilbert space. Dynamical variables are represented by operators operating on the state vector by means of transformations in Hilbert space, but we cannot discuss these generalizations further here.

CHAPTER 2

Rectangular Cartesian Coordinate Systems

2.1 Introduction

In this chapter we shall commence our solution of the various field equations developed in Chapter 1. It is convenient, in terms of the forms of solution obtained, to categorize the field equations according to the symmetry of the coordinate system. Here we shall consider rectangular cartesian solutions in one, two and three dimensions. We shall find it important to express the field equation in coordinates most appropriate to the symmetry of the problem, and failure to do so will generally either mean that we are unable to solve the field equation at all or, in simple cases if a solution *is* obtained, the physical interpretation of the solutions is obscured.

One of our principal means of attack lies in the method of *separation of variables*. There is, of course, no guarantee whatsoever that we will be able to effect a separation of the field equation into a system of single-variable differential equations. Indeed, the number of problems which do conveniently separate in this way is strictly limited, and we shall find in all other cases we shall have to resort to approximate means of solution. It would probably be fair to say that it is this fact more than any other which has prevented the rigorous development of solutions to the various field equations both in classical and quantum mechanical problems.

The expression of the Laplacian operator ∇^2 in rectangular cartesian co-ordinates is given in terms of the spatial partial derivatives as:

$$\nabla^2 \psi(x, y, z) = \frac{\partial^2 \psi}{\partial x^2} + \frac{\partial^2 \psi}{\partial y^2} + \frac{\partial^2 \psi}{\partial z^2} \tag{2.1.1}$$

We may immediately understand this operator as relating to the *curvature* of the generalized potential field $\psi(x, y, z)$, and we will generally attempt to effect a separation of variables by writing

$$\psi(x, y, z) = X(x)Y(y)Z(z) \tag{2.1.2}$$

where X, Y and Z are functions of the single variables x, y and z respectively. For time-dependent field equations we shall attempt to express the normal mode solutions in the form

$$\psi(x, y, z, t) = X(x)Y(y)Z(z)T(t) \tag{2.1.3}$$

In this chapter we shall develop solutions of some of the field equations for one-dimensional systems, or systems which satisfy the field equation in one dimension only and are otherwise constant. As examples of this we might consider the modes of vibration of a flexible stretched string, the propagation of a plane thermal wavefront in a uniform conducting medium, or the propagation of surface waves in a long, narrow trough of liquid.

2.2 Wave Equation for a Uniform Stretched String

As we observed in Chapter 1, the general wave equation is given as (equation (1.4.1))

$$\frac{\partial^2 \psi(x, t)}{\partial x^2} = \frac{1}{c_x} \frac{\partial^2 \psi(x, t)}{\partial t^2} \tag{2.2.1}$$

for a system sustaining wave motion in the x-direction only. c_x represents the velocity of propagation of the wave through the medium in the x-direction, and $\psi(x, t)$ represents the generalized amplitude of the wave at position x and time t. We have already seen that a separation of variables $\psi(x, t) = X(x)T(t)$ effects a separation in terms of the Helmholtz Equation for the spatial wave form $X(x)$ (Section 1.4):

$$\frac{d^2 X(x)}{dx^2} = -k^2 X(x) \tag{2.2.2}$$

In a specific problem such as the modes of vibration of a uniform stretched string we have also to identify the velocity of propagation c_x in terms of the parameters defining the physical state of the string such as its tension, density and so on. We shall now rederive equation (2.2.2), but this time specifically for the present case, and in doing so determine c_x.

Consider the equilibrium of the element of stretched string of length dx and under tension τ (Figure 2.2.1). If the element were perfectly straight there would, of course, be no net force on it. If, however, it is curved we see that a restoring

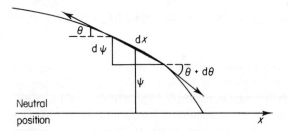

Figure 2.2.1 Geometry for the calculation of the equation of motion of an infinitesimal element of a vibrating string

component develops, and moreover we can see at least qualitatively that the greater the curvature the greater the restoring force. Provided the amplitude ψ of the vibrations is not large and the tension τ remains constant (small θ), then the equation of motion of the element is

$$\rho \ddot{\psi} \, dx = \tau \sin (\theta + d\theta) - \tau \sin \theta \tag{2.2.3}$$

i.e.

$$\ddot{\psi} = \frac{\tau}{\rho} \frac{d\theta}{dx} \qquad (2.2.4)$$

and, again, provided θ is small $\theta \sim d\psi/dx$, we obtain the one-dimensional wave equation

$$\frac{\partial^2 \psi(x, t)}{\partial x^2} = \frac{\rho}{\tau} \frac{\partial^2 \psi(x, t)}{\partial t^2} \qquad (2.2.5)$$

Comparison with equation (2.2.1) enables us to identify the velocity of propagation of the wave as $c = \sqrt{(\tau/\rho)}$. A separation of variables then yields the two equations

$$\frac{d^2 X(x)}{dx^2} = -k^2 X(x) \qquad (2.2.6)$$

and

$$\frac{d^2 T(t)}{dt^2} = -\frac{\tau}{\rho} k^2 T(t) \qquad (2.2.7)$$

where $-k^2$ is the separation constant.

We see from equations (2.2.6) and (2.2.7) that the wave equation (2.2.5) has solutions of the form

$$\psi(x, t) = (A \sin kx + B \cos kx)(C \sin ckt + D \cos ckt) \qquad (2.2.8)$$

where $k = \omega/c$. A linear combination of such solutions also constitutes a solution to the wave equation so we may more generally write

$$\psi(x, t) = \sum_{n=0}^{\infty} (A_n \sin k_n x + B_n \cos k_n x)(C_n \sin ck_n t + D_n \cos ck_n t) \qquad (2.2.9)$$

where the coefficients $A_n, B_n, \dots$, and the integers n may be determined from the boundary conditions, and the initial position and velocity of the string. Equation (2.2.8) represents a *normal mode* solution or *eigenfunction* of the initial wave equation, and equation (2.2.9) is a linear combination of such normal modes. These modes are characterized by a motion in which the amplitude ψ at all points of the vibrating string varies in time in the same way, the frequency and phase being everywhere the same. This is a statement of the independence of the spatial and temporal variables, i.e.

$$\psi(x, t) = X(x)T(t)$$

Suppose now that the string is of length L, its ends being rigidly held such that $\psi(0, t) = \psi(L, t) = 0$ (Dirichlet boundary conditions). Suppose also that the initial form of the string is $\psi(x, 0) = f(x)$, and the initial velocity of the string is $\dot{\psi}(x, 0) = \dot{q}(x)$. The condition $\psi(0, t) = \psi(L, t) = 0$ may be satisfied only if

$B_n = 0$ in equation (2.2.9), except for the trivial non-vibrating case. Our solution now reads

$$\psi(x, t) = \sum_{n=1}^{\infty} A_n \sin(k_n x)\{C_n \sin(ck_n t) + D_n \cos(ck_n t)\} \qquad (2.2.10)$$

where the summation specifically excludes the trivial $(n = 0)$ non-vibrating solution. From the boundary condition at $x = L$, it immediately follows that

$$k_n = \frac{n\pi}{L} \qquad (2.2.11)$$

For the fundamental mode of vibration, $n = 1$, we have $\lambda = L/2$. The boundary condition represents, of course, the physical requirement that the oscillations sustained by the string terminate in nodes at $x = 0, L$. The first few harmonics are shown in Figure 2.2.2. Further application of the boundary conditions yields

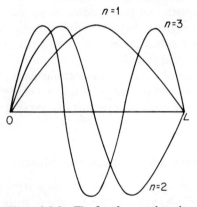

Figure 2.2.2   The first 3 normal modes of vibration of a string fixed at points $x = 0, L$

the remaining coefficients as follows. At $t = 0$ the string adopts the arbitrary functional form $\psi(x, 0) = f(x)$ in the interval $0 \leqslant x \leqslant L$ so that from equation (2.2.10)

$$\psi(x, 0) = f(x) = \sum_{n=1}^{\infty} E_n \sin(k_n x)$$

where $E_n = A_n D_n$. From the orthogonality property of the sine function, multiplication of equation (2.2.11) throughout by $\sin(k_n x)$ and integration yields

$$E_n = \frac{2}{L} \int_0^L f(x) \sin(k_n x)\, dx \qquad (2.2.12)$$

Again, the arbitrary functional form of the initial *velocity* distribution, $\dot\psi(x, 0) = \dot q(x)$ yields

$$\dot\psi(x, 0) = \dot q(x) = \sum_{n=1}^{\infty} k_n c F_n \cos(k_n ct) \qquad (2.2.13)$$

where $F_n = A_n C_n$. Equation (2.2.13) arises, of course, as the first time differential of the general solution, equation (2.2.10). The coefficient F_n may be determined as before so that

$$F_n = \frac{1}{k_n c} \frac{2}{L} \int_0^L \dot{q}(x) \sin (k_n x)\, dx \qquad (2.2.14)$$

and we may now write down the final normal mode solution

$$\psi(x, t) = \sum_{n=1}^{\infty} \sin (k_n x)\{F_n \sin (k_n ct) + E_n \cos (k_n ct)\} \qquad (2.2.15)$$

It is a characteristic of the normal vibrational modes of one-dimensional systems that the higher harmonics are simply related to the fundamental frequency of vibration v_0, by integral multiples. Such simple relationships are not at all common in other vibrating systems, as we shall see, and this accounts for the almost exclusively one-dimensional form, strings, columns of air, etc., for instruments of any musical pretension. A violinist bowing his string can in principle only excite the frequencies given by equation (2.2.15). He can, however, alter the *quality* of the note, that is the accompanying higher harmonics, by bowing closer to the bridge of the violin. What he is effectively doing is changing the functional form of $f(x)$, $\dot{q}(x)$ and consequently, varying the relative amplitudes E_n, F_n.

2.3 Hybridization of the Normal Modes of Vibration

We continue our discussion of the vibration of a stretched uniform string of density ρ, but now consider the case for the string suspended vertically in the gravitational field, the tension increasing as we move up the string. The tension would remain uniform at the value T only in the case of an ideal, massless string. We shall suppose that the tension varies up the string as $T(1 + t(z))$ where $|t| \ll T$ throughout the interval $0 \leqslant z \leqslant L$ (Figure 2.3.1). We might anticipate

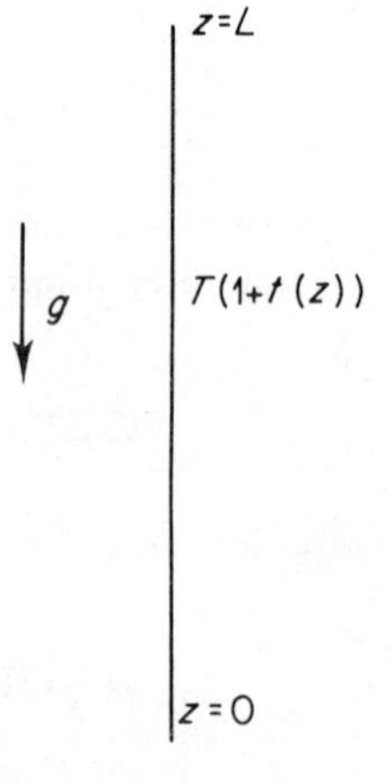

z = L

g

$T(1 + t(z))$

z = 0

Figure 2.3.1 A uniform 'massive' string whose tension varies under gravity as $T(1 + t(z))$

that for such a small perturbation the perturbed normal modes will be very similar to the *unperturbed* ones. Indeed, we may write for the nth perturbed mode

$$\psi_n = A_n\psi_n^0 + \sum_{i \neq n}^{\infty} A_i\psi_i^0 \tag{2.3.1}$$

where ψ_n^0 represents the nth *unperturbed* normal mode and since the perturbation is small we suppose that $A_n \gg A_i$. Again, the frequency of the nth perturbed mode will be simply related to that of the unperturbed mode provided the perturbation is small:

$$\omega_n^2 = (\omega_n^0)^2(1 + \varepsilon_n) \quad |\varepsilon_n| \ll 1 \tag{2.3.2}$$

There is no necessity to define ε_n in terms of the *square* of the vibrational frequency, but as we shall see shortly, this does prove to be a convenient choice. The differential equation describing the nth mode is still of Helmholtz form, but is now written

$$\frac{d^2\psi_n}{dz^2} + \frac{\rho}{T(1 + t(z))}\omega_n^2\psi_n = 0 \tag{2.3.3}$$

where ψ_n represents the perturbed mode. Provided $t(z)$ is small, we may apply the binomial expansion and rewrite equation (2.3.3) as

$$\frac{d^2\psi_n}{dz^2} + \frac{\rho}{T}(1 - t(z))\omega_n^2\psi_n = 0 \tag{2.3.4}$$

Substituting equations (2.3.1) and (2.3.2) into equation (2.3.4) we obtain

$$\left\{ A_n\frac{d^2\psi_n^0}{dz^2} + \sum_{i \neq n}^{\infty} A_i\frac{d^2\psi_i^0}{dz^2} \right\} + \frac{\rho}{T}(1 - t(z))(1 + \varepsilon_n)\omega_n^{0\,2}\left\{ A_n\psi_n + \sum_{i \neq n}^{\infty} A_i\psi_i^0 \right\} = 0 \tag{2.3.5}$$

Neglecting products of small quantities (A_j, $t(z)$, ε_n) and since

$$\frac{d^2\psi_i^0}{dz^2} = -\frac{\rho}{T}\omega_i^{0\,2}\psi_i^0$$

expansion of equation (2.3.5) with some rearrangement yields

$$A_n\left\{ \frac{d^2\psi_n^0}{dz^2} + \frac{\omega_n^{0\,2}}{T}\rho A_n\psi_n \right\}$$
$$+ \frac{\rho}{T}\left\{ \omega_n^{0\,2}(\varepsilon_n - t(z))A_n\psi_n^0 + \sum_{i \neq n}^{\infty} A_i\psi_i^0(\omega_n^{0\,2} - \omega_i^{0\,2}) \right\} = 0 \tag{2.3.6}$$

The expression in the first bracket is zero since it is no more than the equation of motion of the nth normal unperturbed mode. It then follows that

$$\omega_n^{0\,2}[(\varepsilon_n - t(z)]A_n\psi_n^0 + \sum_{i \neq n}^{\infty} A_i\psi_i^0(\omega_n^{0\,2} - \omega_i^{0\,2}) = 0 \tag{2.3.7}$$

We may very simply determine the shift in frequency ε_n of the nth perturbed mode from the nth normal mode by multiplying equation (2.3.7) throughout by $\psi_n^0(z)$ and integrating over the interval $0 \leqslant z \leqslant L$:

$$\omega_n^{0\,2} A_n \left\{ \int_0^L \varepsilon_n \psi_n^0 \psi_n^0 \, dz - \int_0^L t(z) \psi_n^0 \psi_n^0 \, dz \right\}$$
$$+ \sum_{i \neq n}^\infty A_i (\omega_n^{0\,2} - \omega_i^{0\,2}) \int_0^L \psi_i^0 \psi_n^0 \, dz \tag{2.3.8}$$

We should, however, remember that the ψ_i^0 are sine functions and their orthogonality property causes the sum in equation (2.3.8) to vanish since

$$\int_0^L \psi_i^0 \psi_n^0 \, dz = 0$$

We therefore have for the frequency shift of the nth perturbed mode relative to the normal mode of vibration:

$$\frac{\omega_n - \omega_n^0}{\omega_n^0} = \varepsilon_n = \frac{\int_0^L t(z) \psi_n^{0\,2} \, dz}{\int_0^L \psi_n^{0\,2} \, dz} \tag{2.3.9}$$

We see that there is an *increase* in the frequency under these circumstances, and that the shift vanishes as $t(z)$ decreases. Equation (2.3.9) is of very familiar quantum mechanical form when we would say that ε_n was the *expectation value* of the perturbation $t(z)$, the average being taken over the nth normal mode. If the perturbation is large where the amplitude of ψ_n^0 is large then the shift of frequency will be considerable. If the perturbation could somehow be located at the nodes of the normal mode then the perturbation of that normal mode and any other having nodes at these points will suffer no shift in frequency.

To determine the various amplitudes A_i which *hybridize* or contaminate the normal mode under the action of the perturbation, we multiply equation (2.3.7) by $\psi_i^0(z)$ and integrate over the interval $0, L$. Orthogonality of the ψ^0 functions enables us to write

$$-A_n \omega_n^{0\,2} \int_0^L t(z) \psi_n^0 \psi_i^0 \, dz = A_i (\omega_n^{0\,2} - \omega_i^{0\,2}) \int_0^L \psi_i^{0\,2} \, dz \tag{2.3.10}$$

whereupon

$$\frac{A_i}{A_n} = \frac{-\omega_n^{0\,2}}{\omega_n^{0\,2} - \omega_i^{0\,2}} \cdot \frac{\int_0^L t(z) \psi_n^0 \psi_i^0 \, dz}{\int_0^L \psi_i^{0\,2} \, dz} \tag{2.3.11}$$

Notice that the integral on the left-hand side of equation (2.3.10) does not vanish on the grounds of orthogonality owing to the presence of the function $t(z)$ in the integrand. We observe first of all in equation (2.3.11) that hybridization with the *adjacent* normal modes takes place most extensively, the ratio A_i/A_n increasing as $\omega_i^{0\,2} \to \omega_n^{0\,2}$. This proves to be a fundamental principle in the hybridization of normal modes under the action of a perturbation. Generally, hybridization takes place most extensively between normal modes of similar frequency. In the present case they also happen to be the adjacent modes. In some cases when the frequencies are very close, hybridization occurs even in the *absence* of a perturbation. We shall consider some of these cases later. The amplitude A_i of the ith hybridizing normal mode will also be large when $t(z)$, ψ_n^0 and ψ_i^0 are all large in the same spatial region. We are therefore able to describe the nth *perturbed* mode in terms of the nth *unperturbed* normal mode plus a linear combination of the complete set of normal modes, provided the perturbation is small

$$\psi_n = A_n\psi_n^0 + \sum_{i \neq n}^{\infty} A_i\psi_i^0$$

where the hybridization coefficients A_i are given by equation (2.3.11).

Hybridization is a perfectly familiar approach to the discussion of non-normal modes of vibration; the representation of an arbitrary function in terms of a synthesis of eigenfunctions provides a simple example very close to the problem we have been discussing, where the A_i have the significance of Fourier coefficients. The concept of hybridization of normal modes to represent a perturbed normal mode is of outstanding importance in virtually every branch of classical field theory and quantum molecular physics. The number of field equations which can be rigorously solved is very small indeed, and approximate methods must inevitably be employed. The representation in terms of a complete set of 'normal mode' or unperturbed solutions provides a direct and physically appealing route to the final field distribution ψ. As we shall see, the type of *basis function* which forms the complete set of unperturbed solutions need not be a simple trigonometric function, although, as we have seen, the orthogonality of these functions is a useful property. We shall later find that Bessel functions, Legendre functions, Hermite functions and many others constitute adequate basis functions for a hybrid description, depending of course on the nature of the coordinate system.

2.4 The Freely Hanging Chain

Returning to the discussion of Section 2.2, we see that when the tension in a string is a function of position the net restoring force on an element of the string is (cf. equation (2.2.3))

$$\rho\ddot{\psi}\,dx = \tau(x + dx)\sin(\theta + d\theta) - \tau(x)\sin\theta$$

$$= \frac{\partial}{\partial x}\left(\tau(x)\frac{\partial\psi}{\partial x}\right)dx,$$

(2.4.1)

and the resulting wave equation is

$$\frac{\partial}{\partial x}\left(\tau(x)\frac{\partial\psi(x, t)}{\partial x}\right) = \rho\frac{\partial^2\psi(x, t)}{\partial t^2}$$

(2.4.2)

which reduces to the usual wave equation when τ is independent of position. We may now investigate the normal modes of vibration of a freely hanging uniform chain in the gravitational field. If we consider a chain of length l and uniform density ρ suspended at its upper end, and if we take the origin of coordinates at its lower end, then evidently the tension in the chain will vary as

$$\tau(x) = \rho g x$$

If now we assume normal mode solutions of the form

$$\psi(x, t) = X(x)\,e^{-2\pi i v t}$$

where we implicitly require $X(x)$ to be everywhere finite and single-valued, then substitution in the wave equation (2.4.2) yields

$$x\frac{\mathrm{d}^2 X}{\mathrm{d}x^2} + \frac{\mathrm{d}X}{\mathrm{d}x} + \frac{4\pi^2 v^2 X}{g} = 0 \tag{2.4.3}$$

We see immediately that the frequencies of vibration are independent of the density of the chain provided, of course, that it is uniform. This we can understand since an increased density would tend to lower the vibrational frequency, and this would be just compensated by the corresponding increase in tension tending to raise it. To take the analysis further we need to find the eigenfunctions and the associated eigenvalues of equation (2.4.3). This equation is not, however, of any form which we have encountered so far, and so we must anticipate normal mode solutions differing from the simple trigonometric functions which have arisen in earlier sections and it is this which prevents us applying the perturbation approach developed in Section 2.3. Indeed, as it stands, equation (2.4.3) is not of any standard form, although if we make certain substitutions it is possible to transform it into *Bessel's equation*. This equation and its solutions will be discussed at some length in the next chapter: a detailed solution and some properties of the Bessel functions are given in Appendix III.

If we make the substitution $z = \sqrt{(x/l)}$, $X(x) = X(z)$, then it is straightforward to show that equation (2.4.3) becomes

$$\frac{\mathrm{d}^2 X}{\mathrm{d}z^2} + \frac{1}{z}\frac{\mathrm{d}X}{\mathrm{d}z} + \frac{16\pi^2 v^2 l}{g} X = 0 \tag{2.4.4}$$

which, by comparison with equation (III.1) in Appendix III, we see has as its solution the zero-order ($n = 0$) Bessel function:

$$X(z) = J_0(kz) \tag{2.4.5}$$

where $k^2 = 16\pi^2 v^2/g$. The zero-order Bessel function is shown in Figure 2.4.1. As we see, $J_0(y)$ repeatedly passes through zero as its argument increases, and this enables us to satisfy the boundary condition at the upper end of the chain, i.e. $X(l) = 0$. Thus, only those modes of vibration may develop which satisfy this condition, and this enables us to establish the eigenvalues or normal mode

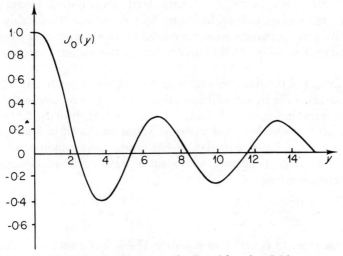

Figure 2.4.1   The zero-order Bessel function $J_0(y)$

frequencies. The top end of the string is at $x = l$, whereupon we have a condition on k such that

$$J_0(k) = 0 \quad (z = 1) \tag{2.4.6}$$

and this determines the normal mode frequencies. If y_j is the value of y in Figure 2.4.1 for which $J_0(y)$ crosses the axis for the jth time, then the eigenfunctions or normal modes of vibration have the form

$$\psi(x, t) = J_0\left(\pi k_j \sqrt{\left[\frac{x}{l}\right]}\right) e^{-2\pi i \nu_j t} \tag{2.4.7}$$

where $\nu_j = (k_j/4)\sqrt{(g/l)}$. The spatial form of the first few normal modes are shown in Figure 2.4.2. If the fundamental mode is of frequency ν_0, then the overtones

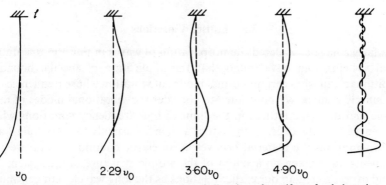

Figure 2.4.2   The first few normal modes of vibration of a uniform freely hanging chain. We see that the normal modes are harmonically unrelated and that for highly excited (high-frequency) normal modes the Bessel functions adopt sinusoidal characteristics

occur at $2.29v_0, 3.60v_0, 4.90v_0, \ldots$. Quite clearly these normal modes are not harmonic, and successive eigenfunctions bear no simple relationship to one another. We can see that each overtone exhibits an extra node, and that these nodes tend to crowd down to the lower end of the chain where the tension is the least.

It is a feature of the Bessel functions in general, and $J_0(y)$ in particular, that for large values of the argument the solutions tend to become sinusoidal. This is already apparent in Figure 2.4.1. Such a situation will apply in the case of highly excited (high frequency) normal modes of vibration shown in Figure 2.4.2. If we reconsider the initial equation (2.4.4) under these conditions, i.e. high frequencies at large argument, we regain a wave equation having simple trigonometric solutions: i.e.

$$\frac{d^2 X}{dz^2} + \frac{1}{z}\frac{dX}{dz} + \frac{16\pi^2 v^2 l}{g}X \to \frac{d^2 X}{dz^2} + \frac{16\pi^2 v^2 l}{g}X \tag{2.4.8}$$

It is important to realize that equation (2.4.4) has a second, independent solution: $Y_0(kz)$. This solution is shown in Appendix III, and as we see, it is rejected on physical grounds in that it does not satisfy the condition that the amplitude of vibration should be everywhere finite. Nonetheless, these functions will recur elsewhere in this book, and we shall defer their discussion till later.

It can be quite easily verified that the spatial eigenfunctions (2.4.7) are mutually orthogonal, that is

$$\int_0^1 J_0(k_i z)J_0(k_j z)z \, dz = 0 \quad i \neq j$$
$$= \tfrac{1}{2}[J_0'(k_j)]^2 \quad i = j \tag{2.4.9}$$

where the prime signifies the first differential with respect to z and k_j are the zeros of $J_0(y)$. Note the weighting function z in the orthogonalization integral (See Section 1.6).

2.5 Lattice Vibrations

Thus far we have considered the propagation of waves in uniform one-dimensional systems, and have found that the application of specific boundary conditions establishes certain normal modes of vibration whose frequencies are very simply related. We now come to consider the vibrational modes of monatomic and diatomic lattices in the limit of high frequency vibrations whose wavelength is of the order of the lattice spacing. Certainly for very long wavelength vibrations the material behaves as an elastic continuum: our previous treatments have taken no account of the atomic structure of the system. We should expect to regain our earlier relations as the long wavelength continuum limit to the present discussion. When, however, the wavelength approaches that of the lattice dimensions we begin to detect features which may be specifically attributed to the periodic nature of the lattice.

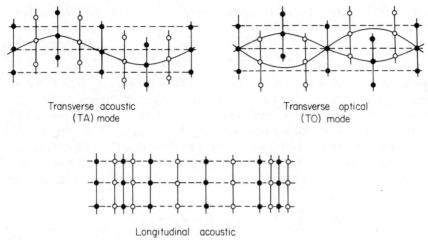

Transverse acoustic
(TA) mode

Transverse optical
(TO) mode

Longitudinal acoustic
(LA) mode

Figure 2.5.1 A comparison of the transverse acoustic (TA), transverse optical (TO) and longitudinal acoustic (LA) modes of lattice vibration. The TO mode is conveniently excited by the incidence of electromagnetic radiation on ionic crystal systems (NaCl, etc.) where alternate planes of atoms are of opposite electrostatic charge. The acoustic modes are those which would be ordinarily excited in the bulk of the crystal by simple mechanical vibration

We consider first the propagation of a travelling wave in the positive x direction in a diatomic lattice. If we arrange that this axis coincides with either the 100, 111 or 110 directions then entire planes of atoms will move in phase as shown in Figure 2.5.1. In the case of diatomic lattices two distinct transverse modes may develop. The transverse acoustic (TA) mode corresponds to a simple transverse waveform as shown in Figure 2.5.1. Such modes may be excited by the propagation of mechanical waves through the lattice, and are often termed *hypersonic modes* since they correspond to very high frequency acoustic excitations (phonons) with a wavenumber k typically $\gtrsim 10^8$ cm^{-1}. The transverse optical (TO) mode, however, frequently arises as an interaction between electromagnetic radiation and the lattice. If the two ionic species constituting the diatomic lattice carry opposite charges as in the case of NaCl, then clearly such optical modes will be excited by the absorption of a photon. A photon will typically have a frequency $\sim 10^{13}$ c.p.s. and the optical phonon will correspondingly have a wavenumber $k \sim 10^3$ cm^{-1}; much lower than the TA modes. The longitudinal modes develop as simple compressional waves within the lattice.

If the spacing of the planes is a then the repeat distance of the diatomic lattice is $2a$. We shall take the mass of the two species as M_1 and M_2, $M_1 > M_2$, and shall take M_1 and M_2 to lie on odd and even planes, respectively.

The motion of, say, the $(2n)$th plane will be strongly coupled to the motion of the neighbouring planes, although we shall first of all restrict the coupling to those immediately adjacent. The force on the $(2n + 1)$th plane is clearly,

$$\begin{aligned}
F_{2n+1} &= C(\psi_{2n+2} - \psi_{2n+1}) - C(\psi_{2n} - \psi_{2n+1}) \\
&= C(\psi_{2n+2} + \psi_{2n-1} - 2\psi_{2n})
\end{aligned} \tag{2.5.1}$$

where we have quite reasonably made the harmonic approximation that the force on plane $2n + 1$ caused by the displacement of planes $2n, 2n + 2$ is proportional to the difference in their displacements. C is the corresponding coefficient of proportionality and will vary from system to system, differing for longitudinal and transverse waves. We shall determine C later for the specific case of the monatomic lattice. Equation (2.5.1) is, of course, a statement of Hooke's Law. The equations of motion of planes $2n$ and $2n + 1$ are, therefore

$$\left.\begin{aligned}
M_1 \frac{d^2\psi_{2n+1}}{dt^2} &= C(\psi_{2n+2} + \psi_{2n} - 2\psi_{2n+1}) \\
M_2 \frac{d^2\psi_{2n}}{dt^2} &= C(\psi_{2n+2} + \psi_{2n-1} - 2\psi_{2n})
\end{aligned}\right\} \tag{2.5.2}$$

We now look for solutions in the form of travelling waves, but with different amplitudes A_{odd}, A_{even} on the odd and even planes:

$$\left.\begin{aligned}
\psi_{2n+1} &= A_{odd} \exp i\{(2n + 1)ka - \omega t\} \\
\psi_{2n} &= A_{even} \exp i\{2nka - \omega t\}
\end{aligned}\right\} \tag{2.5.3}$$

If we now substitute equation (2.5.3) in equation (2.5.2) we immediately obtain

$$\left.\begin{aligned}
-\omega^2 M_1 A_{odd} &= C A_{even}(e^{ika} + e^{-ika}) - 2C A_{odd} \\
-\omega^2 M_2 A_{even} &= C A_{odd}(e^{ika} + e^{-ika}) - 2C A_{even}
\end{aligned}\right\} \tag{2.5.4}$$

Now, this set of equations will only have non-trivial solutions if the determinant of coefficients A_{odd} and A_{even} vanishes, thus

$$\begin{vmatrix} 2C - M_1\omega^2 & -2C\cos ka \\ -2C\cos ka & 2C - M_2\omega^2 \end{vmatrix} = 0$$

Evaluation of the determinant yields:

$$\omega^2 = C\left(\frac{1}{M_1} + \frac{1}{M_2}\right) \pm C\left[\left(\frac{1}{M_1} + \frac{1}{M_2}\right)^2 - \frac{4(1 - \cos^2 ka)}{M_1 M_2}\right]^{\frac{1}{2}} \tag{2.5.5}$$

Equation (2.5.5) represents the *dispersion relation* between frequency and wavelength in the lattice. For an elastic continuum we have the linear dispersion relation $\omega = ck$, where c is the velocity of propagation of the wave in the medium. As $k \to 0$, or alternatively as $\lambda \to \infty$, we should expect to regain the above linear dispersion relation, and as such it represents the continuum limit.

If we first of all consider the case of a monatomic lattice for which $M_1 = M_2 = M$, $A_{odd} = A_{even}$, then equation (2.5.4) (and hence (2.5.5)) simplifies to give

$$\omega^2 = \frac{2C}{M}(1 - \cos ka) = \frac{4C}{M}\sin^2\frac{ka}{2}$$

i.e.

$$\omega = \left(\frac{4C}{M}\right)^{\frac{1}{2}}\left|\sin\frac{ka}{2}\right| \tag{2.5.6}$$

where the $+$ sign of the $\pm$, and the associated optical mode, no longer arise. Equation (2.5.6), then, represents the dispersion relation for a simply-coupled monatomic lattice. This relation is periodic in k-space, reaching its maximum value at $k = \pm\pi/a$, as shown in Figure 2.5.2. The interval $-\pi/a \leqslant k \leqslant \pi/a$ is

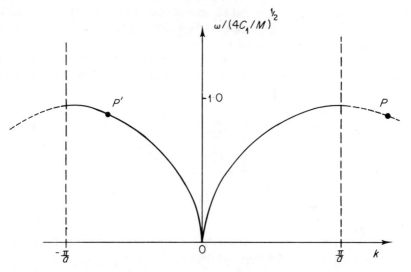

Figure 2.5.2 The dispersion curve for a one-dimensional monatomic lattice of inter-planar spacing a. The dispersion curve replicates in k-space every $k = \pm 2n\pi/a$ so that the points P, P' may be taken as coincident: the region shown here being the first Brillouin zone

termed the *first Brillouin zone*, and is the only zone of interest here, since this range covers all *independent* values of exp (ika). Thus, if we consider a point P on the dispersion curve, then for the purposes of the propagation of lattice vibrations the situation is accounted for within the first Brillouin zone by the point P'. This is a very familiar concept which recurs whenever we consider the propagation of waves in periodic structures. A particularly interesting situation develops at the boundaries of the Brillouin zone when $pk = \pm p\pi/a$, where p is an integer. The general solution of equation (2.5.3) becomes

$$\psi_p = A\,e^{\pm ip\pi}\,e^{-i\omega t} = A\,e^{-i\omega t}\cos p\pi \tag{2.5.7}$$

and this represents a standing wave: $pk = \pm p\pi/a$ is in fact the Bragg condition. Under these conditions waves are not propagated through the lattice. Under the action of an applied electric field, for example, we shall find that the de Broglie waves representing an electron in a periodic lattice exhibit a very similar dispersion relation to the acoustic branch discussed here: we shall return to this point in Section 2.9.

We have made no comment as yet regarding the extent of mechanical coupling between neighbouring atomic planes. Brockhouse *et al.* have compared the experimental dispersion relation for lead at 100 K with a simply modified form of equation (2.5.6):

$$\omega^2 = \frac{2}{M} \sum_0^m C_n(1 - \cos nka) \tag{2.5.8}$$

which takes account of the m adjacent planes on either side. They have calculated the dispersion curve for $m = 5$ and 12, corresponding to coupling with 5 and 12 neighbouring planes. The dispersion relation in the latter case is found to be in excellent agreement with experiment.

The coupling parameters C_n may be very simply determined by the usual method: multiplication of equation (2.5.8) by $\cos(ska)$ throughout and integration over the interval of k yields

$$M \int_{-\pi/a}^{\pi/a} \omega^2(k) \cos(ska) \, dk = 2 \sum_0^m C_n \int_{-\pi/a}^{\pi/a} (1 - \cos nka) \cos(ska) \, dk$$

$$= -2\pi C_s/a$$

i.e.

$$C_s = -\frac{Ma}{2\pi} \int_{-\pi/a}^{\pi/a} \omega^2(k) \cos(ska) \, dk \tag{2.5.9}$$

If we now return to consider the diatomic dispersion relation, equation (2.5.5), we see at small values of the wave number k we have

$$\omega^2 \sim 2C\left(\frac{1}{M_1} + \frac{1}{M_2}\right) \quad \text{optical branch}$$

$$\omega^2 \sim \frac{2C}{M_1 + M_2} k^2 a^2 \quad \text{acoustical branch} \tag{2.5.10}$$

The range of the first Brillouin zone is, in this case, the interval $-\pi/2a \leqslant k \leqslant \pi/2a$ since the repeat distance for the diatomic lattice is $2a$. At $k = \pm\pi/2a$ the roots become

$$\omega^2 = 2C/M_2 \quad \text{optical branch}$$

$$\omega^2 = 2C/M_1 \quad \text{acoustical branch} \tag{2.5.11}$$

The dispersion relation therefore has separate branches for optical and acoustical phonons, and these are shown in Figure 2.5.3. A characteristic feature of the

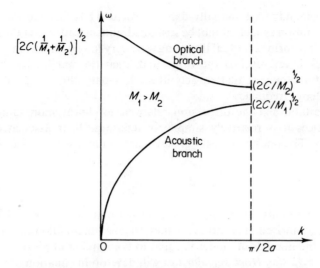

Figure 2.5.3 The optical and acoustic branches of the dispersion curve for a one-dimensional diatomic lattice of period $2a$. A characteristic feature of the propagation of waves in a diatomic lattice is the frequency gap. Solutions at these frequencies represent spatially damped (non-propagating) waves—these might be termed 'Tamm' states from the analogous spatial damping of electronic wavefunctions in a periodic lattice

propagation of waves in a diatomic lattice is the frequency gap. If we look for solutions at these frequencies in the gap, then k turns out to be complex, representing spatially damped waves.

Precisely the same analysis may be given for the weighted string where masses M are distributed along the string with separation a. The dispersion relation is again given by equation (2.5.6), and the maximum frequency which the system can sustain corresponds as before to a wave number $2\pi/\lambda = \pi/a$. However, we may nevertheless attempt to *force* the system to execute higher frequency vibrations. If we allow k to become imaginary and write $k \to ik$ where k is real, then equation (2.5.6) becomes

$$\omega^2 = \frac{2C}{M}[1 - \cos(ika)] \qquad (2.5.12)$$

$$= \frac{2C}{M}[1 + \cosh(ka)] \geqslant \omega^2_{\text{max}} \qquad (2.5.13)$$

The general solution (2.5.3) becomes, under these circumstances,

$$\psi_n = A_n \exp i\{ikna - \omega t\}$$
$$= A_n \exp(-nka) \exp(-i\omega t) \qquad (2.5.14)$$

which corresponds to a spatially damped wave. Thus, beyond the frequency ω_{max}, wave motion as such cannot be sustained, and instead we obtain attenuated solutions of the form (2.5.14). If the string were very long such that the attenuated disturbances never reached the other end, then the weighted string, or the periodic lattice for that matter, acts like a low-pass filter, ω_{max} adopting the role of a 'Debye frequency'.

Having established the dispersion relations for linear monatomic and diatomic lattices it is relatively simple to determine their associated phonon spectra $\sigma(\omega)$. The phonon density of states in the range ω to $\omega + d\omega$ is

$$\sigma(\omega)\,d\omega = n(k)\frac{dk}{d\omega}\,d\omega \qquad (2.5.15)$$

where $n(k)$ represents the number of modes per unit range of k. If the lattice of length L is comprised of N atomic planes, the maximum allowed wavenumber is $N\pi/L$. If the number of modes is equal to the number of particles, it follows that $n(k) = \pi/L$. *Van Hove singularities* will develop in equation (2.5.15) whenever the dispersion curve is horizontal, i.e. when $dk/d\omega \rightarrow \infty$. Some simple analysis confirms the optical and acoustic branches of the phonon spectrum for a diatomic lattice shown in Figure 2.5.4 although such singularities follow directly from the horizontals in the dispersion curve. The physical significance of these singularities is simply that the phonon group velocity, or the phonon effective mass, is zero and amounts to Bragg reflection of the elastic waves within the lattice.

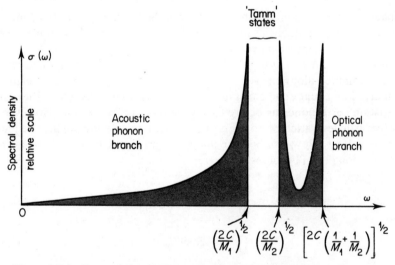

Figure 2.5.4 Spectral density for a one-dimensional diatomic lattice. Van Hove singularities develop at the acoustic upper limiting frequency, and at the upper and lower limiting frequencies on the optical branch. A frequency gap corresponding to spatially damped non-propagating lattice waves separates the acoustic from the optical region

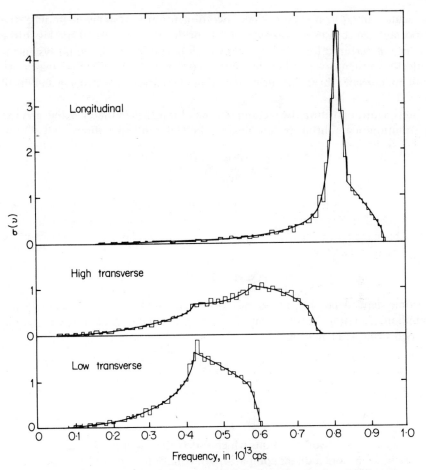

Figure 2.5.5 Anisitropic spectral densities for the aluminium lattice : the histograms are obtained from computed frequencies for 2791 wave vectors. The phonon spectra correspond to the two transverse and one longitudinal mode of propagation of lattice waves, and all branches show evidence of van Hove singularities. (After C. B. Walker, *Phys. Rev.*, **103**, 547 (1956))

Spectral densities for three-dimensional lattices may be determined on the basis of experimental or theoretical dispersion surfaces. The three branches of the aluminium phonon spectrum corresponding to the longitudinal and two transverse polarizations of the lattice wave are shown in Figure 2.5.5.

2.6 Wave Equation for a Uniform Two-Dimensional Membrane

As we shall now see, we may very easily extend our discussion of one-dimensional cartesian systems to include two-dimensional rectangular cartesian membranes. We could possibly anticipate the result by considering the rectangular membrane to be composed of a very large number of vibrating strings. The problem is of particular interest however since two-dimensional systems introduce the important concept of *degeneracy*, and hybridization of normal modes in the absence of perturbation. The logical extension to three-dimensional systems is perfectly straightforward, but introduces no new concept. For that reason we shall not consider three-dimensional rectangular cartesian systems in any detail here.

In its neutral position the rectangular membrane lies in the xy plane, and the distribution of amplitude is described by the Helmholtz equation

$$\frac{\partial^2 \psi}{\partial x^2} + \frac{\partial^2 \psi}{\partial y^2} + \frac{\omega^2}{c^2}\psi = 0 \tag{2.6.1}$$

where $c^2 = \rho/\tau$ is the velocity of propagation of the wave across the membrane, ρ being the density per unit area and τ, the tension, is assumed uniform. In order to establish the normal mode solutions to equation (2.6.1) we assume separability of the variables and write

$$\psi(x, y) = X(x)Y(y) \tag{2.6.2}$$

the time dependence having already been separated off in writing down the Helmholtz equation. Substitution of equation (2.6.2) into equation (2.6.1) yields the two separated equations

$$\frac{d^2 X}{dx^2} + k_1^2 X = 0 \tag{2.6.3}$$

$$\frac{d^2 Y}{dy^2} + k_2^2 Y = 0 \tag{2.6.4}$$

k_1 and k_2 are separation constants which satisfy the condition $k_1^2 + k_2^2 = \omega^2/c^2$, as may be seen from a direct comparison of equations (2.6.3), (2.6.4) and (2.6.1). The solutions to the two separated Helmholtz equations are of the form

$$\left.\begin{aligned} X(x) &= A \sin(k_1 x) + B \cos(k_1 x) \\ Y(y) &= C \sin(k_2 y) + D \cos(k_2 y) \end{aligned}\right\} \tag{2.6.5}$$

where A, B, C, D are arbitrary constants to be determined from the boundary conditions. If the membrane is rigidly fixed at $x = 0$, $y = 0$, then it follows that $B = D = 0$ in equation (2.6.5). Moreover, if the membrane is fixed at $x = a$, $y = b$ the condition that these should also be nodal boundaries gives

$$k_1 = \frac{n\pi}{a}, \qquad k_2 = \frac{m\pi}{b} \tag{2.6.6}$$

where n, m are integers. The normal mode solution is therefore

$$\psi(x, y) = E \sin\left(\frac{n\pi x}{a}\right) \sin\left(\frac{m\pi y}{b}\right) \tag{2.6.7}$$

where $E = AC$. As we have observed, the frequencies of the normal mode solutions are given as

$$\omega = c\sqrt{(k_1^2 + k_2^2)} = \frac{c\pi}{ab}\sqrt{(n^2b^2 + m^2a^2)} \tag{2.6.8}$$

The fundamental mode occurs for $n = m = 1$, and therefore has the frequency

$$\omega_0 = \frac{c}{ab}\sqrt{(a^2 + b^2)} \tag{2.6.9}$$

Clearly there is here no simple harmonic relationship of the type encountered for linear systems. Some slight simplification will occur if the dimensions of the membrane are simply related. In particular we may consider the square membrane for which $a = b$ and the fundamental frequency is

$$\omega_0 = \frac{c}{a}\sqrt{2} \tag{2.6.10}$$

the nodal lines coinciding with the membrane boundary.
Generally,

$$\omega = \frac{c}{a}\sqrt{(n^2 + m^2)} \tag{2.6.11}$$

and the first harmonic occurs for $(n, m) = (1, 2)$ and $(2, 1)$. Again, there is no simple relationship amongst the harmonics. Kauzmann has calculated some of the normal modes for a uniform square membrane, and these are shown in Figure 2.6.1.

Frequency ν_0 $1\cdot58\ \nu_0$ $2\cdot00\ \nu_0$ $2\cdot24\ \nu_0$

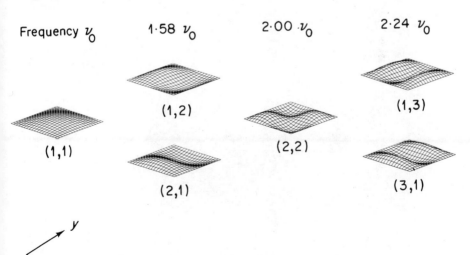

(1,1)   (1,2)   (2,2)   (1,3)

(2,1)   (3,1)

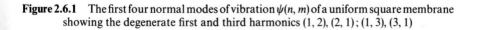

Figure 2.6.1 The first four normal modes of vibration $\psi(n, m)$ of a uniform square membrane showing the degenerate first and third harmonics $(1, 2)$, $(2, 1)$; $(1, 3)$, $(3, 1)$

34

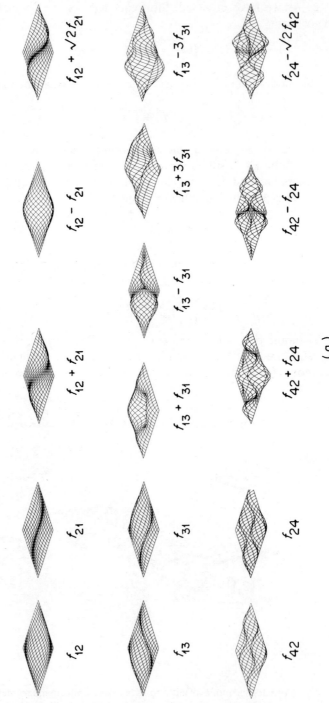

$f_{12} + \sqrt{2}f_{21}$

$f_{13} - 3f_{31}$

$f_{24} - \sqrt{2}f_{42}$

$f_{12} - f_{21}$

$f_{13} + 3f_{31}$

$f_{42} - f_{24}$

$f_{13} - f_{31}$

$f_{12} + f_{21}$

$f_{13} + f_{31}$

$f_{42} + f_{24}$

f_{21}

f_{31}

f_{24}

f_{12}

f_{13}

f_{42}

(a)

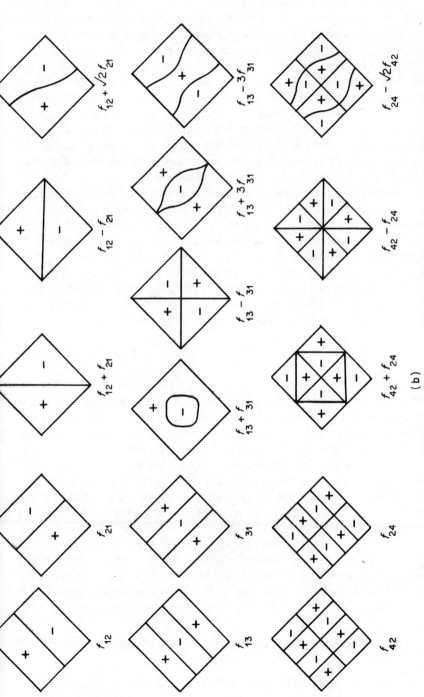

Figure 2.6.2 Hybridization of some degenerate normal modes $\psi(n, m)$ of a square membrane. The $+(-)$ sign indicates that the membrane is displaced above (below) the equilibrium position: the boundary between the two regions being defined by the stationary nodal lines. (W. Kauzmann, *Quantum Chemistry* 1957. Reproduced by permission of Academic Press, Inc.)

(b)

The frequency of the second harmonic is $c\sqrt{5}/a = 1.58\omega$. The mode $(2, 2)$ is seen to have a frequency twice that of the fundamental, whilst the normal modes $(1, 3)$ $(3, 1)$ have identical frequencies at $2.24\omega_0$. This is an example of *double degeneracy*, that is, two distinct normal modes having the same frequency. Clearly most of the modes will be doubly degenerate and this arises from the symmetry of the problem. If the symmetry were removed, and we reconsidered the rectangular membrane, we see from equation (2.6.8) that the normal (rectangular) modes $(1, 2)$ $(2, 1)$ are not degenerate, nor indeed are the general normal modes (n, m) (m, n) in this case. We might anticipate that for stationary vibrations in a cubic cavity we should encounter triply degenerate normal modes arising from the permutation of (l, m, n), whilst any departure from symmetry would generally result in the destruction of degeneracy.

Superposition of degenerate normal modes results in a new normal mode, and this is a particularly important property of the degenerate solutions. Such a linear combination of degenerate modes does, of course, constitute hybridization although in the absence of perturbation. We recall that the extent of hybridization amongst the normal mode solutions depends upon the closeness of the frequencies. In the degenerate case, of course, they are identical and hybridization may be recognized as an alternative choice of orthonormal eigenfunctions for a degenerate eigenvalue. Clearly the two degenerate normal mode solutions (m, n), (n, m) for the square membrane may be combined to give the hybrid

$$\psi(x, y) = \left\{ P \sin\left(\frac{n\pi x}{a}\right) \sin\left(\frac{m\pi y}{a}\right) + Q \sin\left(\frac{m\pi x}{a}\right) \sin\left(\frac{m\pi y}{a}\right) \right\} \quad (2.6.12)$$

We see that this is a normal mode solution which satisfies the original Helmholtz equations (2.6.3), (2.6.4). P and Q are arbitrary constants, and Kauzmann has calculated some of the degenerate hybrid modes for a square membrane. These are shown in Figure 2.6.2. We conclude, therefore, that hybridization occurs particularly easily amongst degenerate modes, and less extensively as the difference in frequency of the two modes increases. We formed this conclusion earlier in considering the normal modes of a heavy vertical string, Section 2.3: equation (2.3.11). Equation (2.6.12) represents a normal mode solution to the Helmholtz equation, regardless of the values of the coefficients P and Q. It would therefore appear possible to construct an infinite number of normal mode solutions (n, m) all having the same frequency $c\sqrt{(n^2 + m^2)}/a$. Not all of these modes will, however, be independent. The degeneracy of the mode is in fact defined as the number of *linearly independent* normal modes of the same frequency. If we can construct any one of these normal mode solutions from a suitable combination of the others, then clearly it is not an independent solution.

2.7 The Schrödinger Wave Equation: Solutions for Which the Potential is a Constant

Both the particle and wavelike aspects of electromagnetic radiation have, at various stages in the development of the subject, been exploited in the description of the interaction between radiation and matter. It is interesting that it was Newton who first proposed a corpuscular hypothesis for the rectilinear propagation of light, but he had some difficulty in accounting for diffraction and interference effects. The wave theory of light was later extensively developed by Huyghens, Young, Fresnel and Fraunhofer, and highly satisfactory explanations of diffraction phenomena were established on this basis. Einstein, however, revived the corpuscular aspect of the radiation field in his description of the photo-electric effect in 1905. Of course we now accept the *wave–particle duality*, and now understand the radiation to propagate in the form of *wave packets*, and as such simultaneously possess wave- and particle-like properties which are inextricably related.

De Broglie hypothesized that there might similarly be some wavelength associated with matter, and proposed the relation

$$\lambda = \frac{h}{p} \tag{2.7.1}$$

where λ is the de Broglie wavelength of the matter wave, h is Planck's constant and p is the particle momentum. De Broglie pointed out that photons in the Compton effect have an apparent mass $m = h\nu/c^2$ where ν is the radiation frequency and c its velocity of propagation. It follows directly that the photons have a wavelength given by equation (2.7.1). De Broglie initially proposed relation (2.7.1) to account for the stability of the Bohr orbits of the hydrogen atom. The basis of the stability was previously somewhat *ad hoc*, and de Broglie's contribution was to show that the circumference of each stationary (stable) Bohr orbit was an integral multiple of the de Broglie wavelength $\lambda = h/p$. De Broglie was effectively suggesting that the stable Bohr orbits represented standing wave forms of some sort.

The existence of a wavelength and of interference phenomena of the type proposed for electrons was soon experimentally verified in the electron diffraction experiments of Davisson and Germer, and G. P. Thomson. These phenomena are not of course restricted to electrons, but apply to all particles microscopic and macroscopic, although in the latter case the associated wavelength is so small that the wave packets behave essentially as particles and the wavelike aspects are not at all apparent.

We now extend the wave equation

$$\nabla^2 \psi = \frac{1}{c^2} \frac{\partial^2 \psi}{\partial t^2}$$

to include the de Broglie waves. It follows from (2.7.1) that the wavelength associated with a particle of kinetic energy K is

$$\lambda^2 = h^2/2mK \tag{2.7.2}$$

and on separating the variables the Helmholtz equation may be written

$$\nabla^2 \psi + \frac{4\pi^2}{\lambda^2} \psi = 0 \tag{2.7.3}$$

where ψ is the spatial (time independent) wavefunction. From equation (2.7.2) we may directly write

$$\nabla^2\psi + \frac{8\pi^2 m}{h^2}(E - V)\psi = 0; \quad E - V = K \qquad (2.7.4)$$

where E is the *total* energy and is assumed constant, and V is the potential energy of the particle, generally a function of position. Equation (2.7.4) is *Schrödinger's equation*, and the precise form of the wavefunction ψ will, of course, depend on the detailed form of the potential function V. We shall now consider some one-dimensional solutions to the Schrödinger equation for various forms of the potential function, $V(x)$.

Consider first the free particle case, for which $V(x) = 0$. The corresponding wave equation has the form

$$\frac{d^2\psi(x)}{dx^2} + \frac{8\pi^2 mE}{h^2}\psi(x) = 0 \qquad (2.7.5)$$

and has oscillatory standing wave solutions of the form

$$\psi(x) = A \sin kx + B \cos kx \qquad (2.7.6)$$

where A and B are arbitrary constants and k^2 (the wave number) $= 8\pi^2 mE/h$. Since there are no boundary conditions all the constants including k may take on any value, and there is no quantization. Moreover, since the solution extends over the region $x = \pm\infty$ the wavefunction (2.7.6) cannot be normalized: the normalization integral becomes

$$\int \psi_k(x)\psi_k^*(x)\, d\tau = \int d\tau$$

and this increases without limit with τ. We may, however, normalize the uniform particle *density* by writing.

$$\int_x^{x+L} \psi_k(x)\psi_k^*(x)\, dx = \rho L$$

where $L \gg 2\pi/k$, and ρ is the particle density. Certainly this means of normalization is justified from a physical point of view: a uniform probability distribution is physically meaningful and so the associated plane wavefunction should be considered acceptable. The energy–wavenumber relation is shown in Figure 2.7.1 and we shall find that this function changes in a characteristic fashion as soon as we apply boundary conditions: quantization of the energy levels occurs.

Suppose the particle now moves in a one-dimensional well, and the motion is restricted to the region $x = 0$, $x = a$ by regions of infinitely high potential. Such a situation holds for conduction electrons in a metal confined by the surface boundaries. Outside the potential well, that is for $0 \geqslant x \geqslant a$, the stationary one-dimensional Schrödinger equation (2.7.4) becomes

$$\frac{d^2\psi}{dx^2} - \infty\psi = 0 \qquad (2.7.7)$$

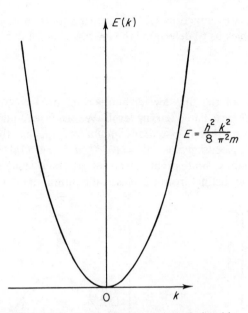

Figure 2.7.1 Energy–wave number relationship $E(k)$ for a free particle. The absence of any boundary conditions ensures a continuous distribution of energy states: no quantization occurs. This curve is identical to the classical energy–momentum relationship

which can be satisfied only if ψ is zero in the region $0 \geqslant x \geqslant a$, outside the well. Within the interval $0 \leqslant x \leqslant a$, that is within the well, the potential is zero, and this corresponds to the free-particle case. The Schrödinger equation takes the form

$$\frac{d^2\psi}{dx^2} + \frac{8\pi^2 mE\psi}{h^2} = 0 \qquad (2.7.8)$$

and this again has solutions of the form

$$\psi = A \sin kx + B \cos kx \qquad (2.7.9)$$

where A and B are constants and $k^2 = 8\pi^2 mE/h^2$. Unlike the free-particle case, however, we now have the specific boundary condition $\psi = 0$ at $x = 0$. This leads us to set $B = 0$ and adopt as the final solution

$$\psi = A \sin kx \qquad (2.7.10)$$

The second boundary condition, $\psi(a) = 0$, places a restriction on the values of k which may be adopted by the wavefunction. Clearly, the condition

$$k = \frac{n\pi}{a} \qquad (2.7.11)$$

40

ensures that the wavefunction (2.7.10) satisfies both boundary conditions. Moreover, it enables us to determine the energy levels of the system as

$$E = \frac{n^2 h^2}{8ma^2} \tag{2.7.12}$$

The spatial form of the first few eigenfunctions are shown in Figure 2.7.2 together with their associated energy levels. We see from equation (2.7.12) that the energy of the particle is essentially *quantized*, and that the spacing of the energy levels is exceedingly close, $\sim(10^{-54}/ma^2)$ erg. Only for particles of exceedingly small mass, and potential wells of extremely small dimension, is the energy distribution distinct from the classical continuum of energy states.

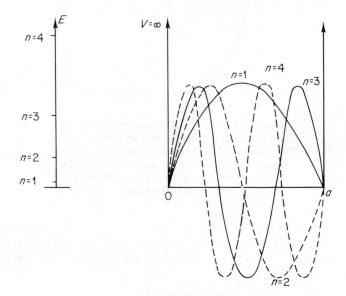

Figure 2.7.2 The spatial form of the first few eigenfunctions and their associated energy eigenvalues for a particle constrained to the interval $0 < x < a$ by regions of infinitely high potential. The quantization has arisen directly from the application of boundary conditions at $x = 0, a$

Classically, of course, we would say that the particle had an equal probability of being located at any point on the interval $0 \leqslant x \leqslant a$. However, the quantum mechanical distribution differs from this, the spatial probability distribution for the nth state being

$$|\psi(x)|^2 = A^2 \sin^2\left(\frac{n\pi x}{a}\right) \tag{2.7.13}$$

corresponding to standing probability waves of de Broglie wavelength $\lambda = 2a/n$. The probability density $|\psi(x)|^2$ varies between 0 and A^2, where A may be determined from the normalization condition

$$1 = A^2 \int_0^a \sin^2 \left(\frac{n\pi x}{a} \right) dx \qquad (2.7.14)$$

since the particle must be located somewhere within the interval $x = 0, a$. From the de Broglie relation $\lambda = h/p$, where p is the particle momentum, we see that as the momentum or energy of the particle increases so the oscillations in $|\psi|^2$ crowd together, and the distribution of probability approaches the classical limit (Figure 2.7.3). This passage to the classical limit at large values of n or in highly excited (large n) states was noted by Bohr, and is termed the *Bohr correspondence principle*.

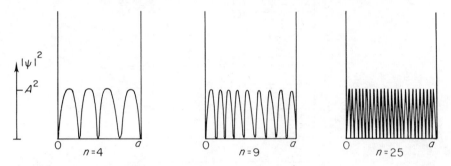

Figure 2.7.3 The probability distribution $|\psi_n|^2$ for a particle in states $n = 4, 9$ and 25 constrained by regions of infinitely high potential to move on the interval $0 < x < a$. We see that the probability distribution tends to the uniform classical distribution as $n \to \infty$

The eigenfunctions for a three-dimensional rectangular well of dimension $a \times b \times c$ surrounded by a region of infinitely high potential may be very easily shown to be of the form

$$\psi(x, y, z) = A \sin \left(\frac{n_1 \pi x}{a} \right) \sin \left(\frac{n_2 \pi y}{b} \right) \sin \left(\frac{n_3 \pi z}{c} \right) \qquad (2.7.15)$$

where as before

$$E = \frac{h^2}{8m} \left[\left(\frac{n_1}{a} \right)^2 + \left(\frac{n_2}{b} \right)^2 + \left(\frac{n_3}{c} \right)^2 \right] \qquad (2.7.16)$$

As we observed in Section 2.6, such solutions are liable to show extensive *degeneracy*, particularly if the dimensions of the well, a, b, c bear simple numerical relationships to one another. If the well is cubical of side a, then extensive degeneracy develops, principally in the form of triply degenerate states.

In Figure 2.7.4 we enumerate some of these states and their eigenvalues, together with their associated degeneracies. This extensive degeneracy arises because of the high degree of symmetry in the potential well. If we were to very slightly modify the potential within the well by amounts $\delta V_x, \delta V_y, \delta V_z$ or adjust

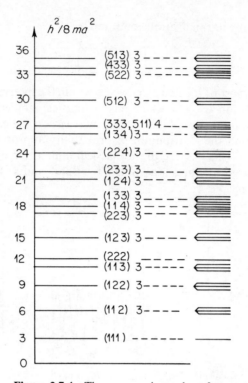

Figure 2.7.4 The energy eigenvalues for a particle constrained to move in a cubical box of side a. These states, by virtue of the cubical symmetry of the constraint, are highly degenerate. The extent of the degeneracy is shown schematically on the right: a slight disturbance of the symmetry reveals (at least partially) the degeneracy of the eigenstate

the dimensions of the well by amounts $\delta x, \delta y, \delta z$ then we should obtain a *splitting* of the degenerate states, and the degree of degeneracy would become apparent. The perturbation is not sufficient to appreciably modify the eigenfunctions, but does have the effect of separating out the energy levels as shown in Figure 2.7.4. Such a situation does occur of course in the Zeeman and Stark effects when degenerate quantum states are separated by the application of an external magnetic or electric field (Section 4.10).

The situation becomes somewhat more complicated if the potential well is of finite depth V_0 as, for example in Figure 2.7.6. In region II ($0 \geqslant x \geqslant a$) the potential is zero, as before. But in regions I and III the potential is V_0. Two distinct situations may be identified according as $E \gtrless V_0$. If the total energy $E > V_0$ then we have the free particle case, and solutions to the Schrödinger equation will be oscillatory in all three regions, although the amplitude of the oscillations in ψ, and correspondingly of the probability density $|\psi|^2$, will be greater in the interval $0 \leqslant x \leqslant a$. Classically the effect on a particle of total energy E passing a region of low potential is to increase its velocity: quantum mechanically there will be a decrease in the de Broglie wavelength. The bound particle case corresponding to $E < V_0$ results in an oscillatory solution in region II, but spatially damped solutions in regions I and III. This is, of course, contrary to classical experience, according to which the particle is constrained to remain within the region $x = 0, a$.

We shall first consider the bound particle case $(E < V_0)$.

Solutions to the Schrödinger equation in the three regions may be immediately written down:

$$\psi_{\text{I}} = A_1 \, e^{ik_1 x} + B_1 \, e^{-ik_1 x} \qquad (2.7.17)$$

$$\psi_{\text{II}} = A_2 \, e^{ik_2 x} + B_2 \, e^{-ik_2 x} \qquad (2.7.18)$$

$$\psi_{\text{III}} = A_3 \, e^{ik_1 x} + B_3 \, e^{-ik_1 x} \qquad (2.7.19)$$

where $k_1 = 2\sqrt{[2\pi(E - V_0)]}/h, k_2 = 2\sqrt{(2\pi E)}/h$.

We see that in regions I and III the condition that the wavefunction does not diverge at $x = \pm\infty$ requires that $B_1 = A_3 = 0$, the probability amplitude otherwise increasing indefinitely with x. In the bound particle case k_1 is imaginary and so solutions ψ_{I} and ψ_{III} are spatially damped functions outside the potential well. The solutions inside the well, region II, correspond to purely oscillatory functions, k_2 being a real positive quantity. Our solutions are therefore of the form

$$\psi_{\text{I}} = A_1 \, e^{ik_1 x} \qquad (2.7.20)$$

$$\psi_{\text{II}} = A_2 \, e^{ik_2 x} + B_2 \, e^{-ik_2 x} \qquad (2.7.21)$$

$$\psi_{\text{III}} = B_3 \, e^{-ik_1 x} \qquad (2.7.22)$$

Since the wavefunction must be everywhere continuous, it remains to adjust the various coefficients so that the amplitudes and gradients at $x = 0, x = a$ are continuous. These conditions ensure that the wavefunction varies smoothly from one region to the next. Thus we require at $x = 0$:

$$\psi_{\text{I}} = \psi_{\text{II}}$$

$$\frac{\mathrm{d}\psi_{\text{I}}}{\mathrm{d}x} = \frac{\mathrm{d}\psi_{\text{II}}}{\mathrm{d}x}$$

i.e.

$$\left. \begin{array}{l} A_1 = A_2 + B_2 \\ k_1 A_1 = k_2 A_2 - k_2 B_2 \end{array} \right\} \qquad (2.7.23)$$

and at $x = a$

$$\left. \begin{array}{l} B_3 \, e^{-ik_1 a} = A_2 \, e^{ik_2 a} + B_2 \, e^{-ik_2 a} \\ -k_1 B_3 \, e^{-ik_1 a} = k_2 A_2 \, e^{ik_2 a} - k_2 B_2 \, e^{-ik_2 a} \end{array} \right\} \qquad (2.7.24)$$

Eliminating A_1 between equations (2.7.23) and B_3 between equations (2.7.24) we obtain

$$(k_1 - k_2)A_2 + (k_1 + k_2)B_2 = 0$$

$$(k_1 + k_2)A_2 e^{ik_2a} + (k_1 - k_2)B_2 e^{-ik_2a} = 0$$

These equations have a non-trivial solution only if the determinant of the coefficients of A_2 and B_2 vanishes:

$$\begin{vmatrix} k_1 - k_2 & k_1 + k_2 \\ (k_1 + k_2)e^{ik_2a} & (k_1 - k_2)e^{-ik_2a} \end{vmatrix} = 0 \qquad (2.7.25)$$

i.e.

$$(k_1 - k_2)^2 e^{-ik_2a} - (k_1 + k_2)^2 e^{ik_2a} = 0 \qquad (2.7.26)$$

Remembering that $k_1 = \sqrt{[2m(E - V_0)]}/\hbar$ is imaginary and $k_2 = \sqrt{(2mE)}/\hbar$ is real, equation (2.7.26) may be written in transcendental form:

$$\cot \frac{a}{\hbar}\sqrt{(2mE)} \quad \text{or} \quad -\tan\frac{a}{\hbar}\sqrt{(2mE)} = \sqrt{\left|\frac{E}{V_0 - E}\right|} \qquad (2.7.27)$$

The roots of this equation may be determined graphically by finding the points of intersection of the tan and cot functions with $\sqrt{[E/(V_0 - E)]}$, and these are shown schematically in Figure 2.7.5.

The cot condition in equation (2.7.27) leads to solutions of the form

$$\psi_1 = 2A_2 \cos\left[a\sqrt{(2mE)}/\hbar\right] \exp\left[(\sqrt{[2m(E - V_0)]}/\hbar)(x + a)\right]$$

$$\psi_{II} = 2A_2 \cos\left[a\sqrt{(2mE)}/\hbar\right]$$

$$\psi_{III} = 2A_2 \cos\left[a\sqrt{(2mE)}/\hbar\right] \exp\left[(-\sqrt{[2m(E - V_0)]}/\hbar)(x - a)\right]$$

where the coefficient A is to be determined by normalization. Such solutions are said to be *symmetric*, $\psi(x) = \psi(-x)$, and have even parity. The lowest two energy eigenfunctions of this type are shown in Figure 2.7.6(a). Higher energy functions will be more oscillatory but will remain symmetric. The energy levels associated with these even parity solutions are given as the intersection of the cot curves with the family $\sqrt{[E/(V_0 - E)]}$. In Figure 2.7.5 we show a family of such curves for various values of the well depth V_0 for fixed E. The curves a through d represent decreasing values of the well depth. We see that as the well depth decreases the number of intersections with the cot curves decreases: nevertheless, there is always at least one bound state as $V_0 \to 0$.

The tan condition in equation (2.7.27) leads to solutions of the form

$$\psi_1 = 2iA_2 \sin\left[a\sqrt{(2mE)}/\hbar\right] \exp\left[(\sqrt{[2m(E - V_0)]}/\hbar)(x + a)\right]$$

$$\psi_{II} = 2iA_2 \sin\left[a\sqrt{(2mE)}/\hbar\right]$$

$$\psi_{III} = 2iA_2 \sin\left[a\sqrt{(2mE)}/\hbar\right] \exp\left[(-\sqrt{[2m(E - V_0)]}/\hbar)(x - a)\right]$$

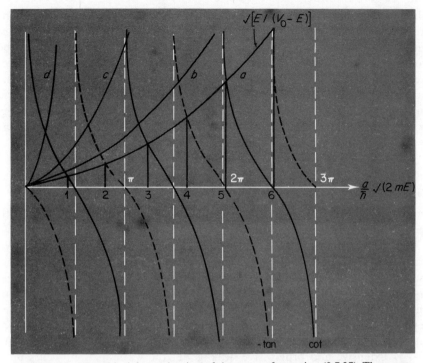

Figure 2.7.5 Graphical determination of the roots of equation (2.7.27). The curves a through d represent the family $\sqrt{[E/(V_0 - E)]}$ for decreasing values of the well depth V_0. The intersections with the cot (full line) curves represent *symmetric* bound eigenstates, whilst the intersections with the tan (broken line) curves represent *antisymmetric* bound eigenstates. We see that symmetric bound 'cot' states remain as the well depth $V_0 \to 0$, whilst no bound 'tan' states exist for $a\sqrt{(2mE)}/\hbar < \pi/2$

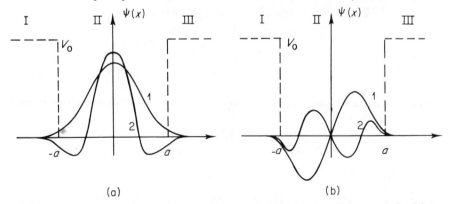

Figure 2.7.6 (a) The lowest two symmetric bound states for a particle constrained by a potential V_0 at $x = \pm a$. (b) The lowest two antisymmetric bound states for a particle constrained by the potential V_0 at $x = \pm a$. If $E < V_0$, corresponding to the bound particle case shown here, then the regions I and III are classically forbidden. Quantum mechanically we see that there is a finite probability of finding the particle in the classically forbidden regions: this 'leakage' or 'tunnelling' decreases with increasing V_0

where, again, A_2 is to be determined by normalization. These wavefunctions are *antisymmetric*, $\psi(x) = -\psi(-x)$, and are of odd parity. The eigenvalues are determined this time by the intersections with the $-\tan$ curve. We see that as the well-depth decreases the number of intersections with the tan functions decreases until for $V_0 < \hbar^2\pi^2/8ma^2$ no intersections occur, and for this value of the parameters a and V_0, no stable bound states may develop for antisymmetric functions. As $V_0 \to \infty$ so the family of curves become coincident with the energy axis and since this axis is in units of $\hbar^2/2ma^2$, we see that we regain the solutions to the infinite well problem (equation (2.7.12))

$$\text{(symmetric)} \quad E_{2n+1} = (2n+1)^2\pi^2\hbar^2/8ma^2$$

$$\text{(antisymmetric)} \quad E_{2n} = (2n)^2\pi^2\hbar^2/8ma^2$$

The symmetric functions reduce to pure cosine solutions, whilst the antisymmetric functions reduce to pure sine form.

The spatial form of the first few eigenfunctions is shown in Figure 2.7.6. We see that apart from developing the characteristic oscillatory form, there exists the finite probability of finding the particle located in the classically inaccessible regions I and III. This phenomenon is termed *quantum tunnelling* and is of purely quantum mechanical origin. The distance the particle penetrates depends particularly upon the magnitude of the function $k_1 = \sqrt{[2m(E - V_0)]}/\hbar$, that is, the height of the potential barrier V_0 relative to the total energy E. As $E \to V_0$ extensive and progressive 'leakage' of the wavefunction occurs until $E = V_0$, after which the solutions in all three regions become oscillatory. This corresponds to the unbound particle case $(E > V_0)$ where we again have the solutions (2.7.17), (2.7.18), (2.7.19). Now, however, we have no reason to discard any of the coefficients since the solutions are everywhere well-behaved. Let us, however, consider a particle of total energy $E > V_0$ moving to the right as shown in Figure 2.7.7. Then in region III the solution may be written

$$\psi_{\text{III}} = A_3\,e^{ik_1x}$$

whilst in region II

$$\psi_{\text{II}} = A_2\,e^{ik_2x} + B_2\,e^{+ik_2x}$$

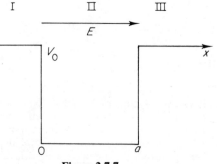

Figure 2.7.7

Continuity of the wavefunction ψ and its derivative $d\psi/dx$ at a requires that

$$A_3 e^{ik_1 a} = A_2 e^{ik_2 a} + B_2 e^{-ik_2 a}$$

$$\frac{k_1}{k_2} A_3 e^{ik_1 a} = A_2 e^{ik_2 a} - B_2 e^{-ik_2 a}$$

whereupon

$$A_2 = \tfrac{1}{2}\left(1 + \frac{k_1}{k_2}\right) A_3 e^{i(k_1 - k_2)a}$$

$$B_2 = \tfrac{1}{2}\left(1 - \frac{k_1}{k_2}\right) A_3 e^{i(k_1 + k_2)a}$$

(2.7.28)

Clearly, provided A_3 is non-zero, $k_1 \neq k_2$ and both k_1 and k_2 are positive, then the coefficients A_2 and B_2 cannot vanish. In particular there is a *reflected* wave in region II whose maximum probability density is $|B_2|^2$. This partial reflection quite clearly is a non-classical result, although such a phenomenon does occur for light rays at the boundary between two regions of differing refractive index.

Notice that these continuity relations (2.7.28) can be satisfied for all values of the energy (provided $E > V_0$) and the energy levels are therefore not quantized. The coefficients $|A_1|^2, |A_2|^2$ and $|A_3|^2$ are all proportional to the 'forward' beam intensities, whilst $|B_2|^2$ is proportional to the reflected beam intensity. In the case of a rectangular potential barrier of height $V_0 > E$, we should expect waveforms of the type shown schematically in Figure 2.7.8. The incident wave

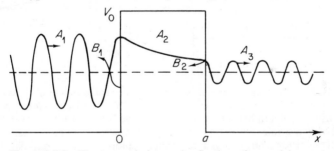

Figure 2.7.8   The transmission and reflection of a particle at a finite potential barrier

amplitude A_1 gives rise to a partially reflected wave B_1 at $x = 0$. The forward wave decays within the barrier to partially reflect at $x = a$, and partially transmit giving a forward wave of amplitude A_3. The transmission coefficient T for the barrier is therefore

$$T = |A_3|^2/|A_1|^2$$

(2.7.29)

and it may be shown quite easily that for the case of the rectangular barrier shown in Figure 2.7.8

$$T = \left\{ 1 + \frac{V_0^2 \sinh^2 \sqrt{[2m(V_0 - E)a^2]}/\hbar}{4E(V_0 - E)} \right\}^{-1}$$

(2.7.30)

In the limit known as the *Born Approximation*: $2m(V - E_0)a^2/\hbar^2 \gg 1$ this reduces to

$$T \sim \frac{16E(V_0 - E)}{V_0^2} \exp\left(-2\sqrt{[2m(V_0 - E)a^2]}/\hbar\right) \qquad (2.7.31)$$

where the exponent is termed the *Gamow factor*.

There are many examples of quantum tunnelling, particularly in the case of electrons whose low mass strongly influences the size of the transmission co-efficient. Cold emission of electrons from metal surfaces in the vicinity of high potential gradients is effectively controlled by the tunelling process. Another important example occurs in the case of α-particle emission from the nucleus. Classically an α-particle of energy $E < V_0$ is confined to the central well of the nuclear potential by the attractive forces (Figure 2.7.9). At some distance from

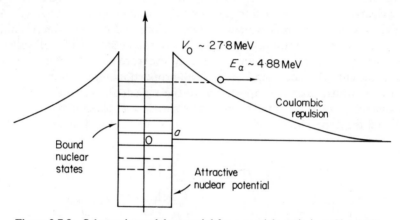

Figure 2.7.9 Schematic model potential for α-particle emission. Classically, α-particles must possess an energy >27.8 MeV in order to surmount the repulsive Coulomb barrier and enter the attractive core. In fact, particles of 4·88 MeV are able to penetrate the barrier and occupy bound nuclear states. Such α-particle emission and resonance capture is, of course, classically forbidden, and provides convincing support for the quantum tunnelling phenomenon

the well the potential is a repulsive Coulombic function, and to escape the α-particle would classically have to surmount the barrier. Taking the radius of the radium nucleus to be 9.1×10^{-13} cm the Coulombic energy of an α-particle just outside the nucleus is 27·8 MeV. The emission of an α-particle from an internal bound state with an energy of 4.88 MeV is classically forbidden. There is also a problem of causality. The particles are identical entities, yet there is no classical explanation as to why certain particles should be spontaneously emitted at an early stage whilst others remain within the nucleus for several half lifetimes (radium half-life = 1620 years). Of course, the phenomenon of quantum tunnelling explains the appearance of particles in classically forbidden regions whilst the essentially indeterministic, probabilistic behaviour of a particle eliminates any question of causality: all we can state is a probability of spontaneous emission occurring. This probability is given as the product of frequency of bombardment times the transmission probability, and this is a constant in time.

The success of the quantum mechanical account of α-decay is undoubtedly one of the major triumphs of wave mechanics.

2.8 The Linear Harmonic Oscillator

In this section we shall consider solutions to the Schrödinger equation for the more general problem in which the potential $V(\mathbf{r})$ is a function of position. Generally, given spatial forms of the potential function characterize the specific functional forms of solution to the Schrödinger equation.

A particularly important system corresponds to the quantized vibration of a particle in the harmonic potential $V(x) = \frac{1}{2}kx^2$ where k is a coefficient of proportionality. It is instructive to first consider the problem classically. The force acting on the particle is $-dV/dx = kx$ and the equation of motion for the particle of mass m follows as

$$m\frac{d^2x}{dt^2} = -kx \qquad (2.8.1)$$

which has the general solution

$$x(t) = A \sin\left[\sqrt{\left(\frac{k}{m}\right)}t + \phi\right] \qquad (2.8.2)$$

where A represents the maximum amplitude of vibration and ϕ is an arbitrary phase constant. The displacement passes through a sinusoidal cycle in a period $\tau = 2\pi\sqrt{(m/k)}$. The velocity of the particle

$$\dot{x}(t) = A\sqrt{\left(\frac{k}{m}\right)}\cos\left[\sqrt{\left(\frac{k}{m}\right)}t + \phi\right] \qquad (2.8.3)$$

is a maximum as it passes through the centre of oscillation when its energy is entirely kinetic, and is zero at $x = \pm A$ when the energy is purely potential. The total energy is $kA^2/2$ which is positive, real, and can adopt any value. Classically the oscillator cannot ever be found outside the spatial interval $-A \leqslant x \leqslant A$, and oscillations are restricted to lie within this range.

Quantum mechanically the equation governing the distribution of probability amplitude, ψ, would be

$$\frac{d^2\psi}{dx^2} + \frac{8m\pi^2}{h^2}(E - \frac{1}{2}kx^2)\psi = 0 \qquad (2.8.4)$$

where E is the total energy of the oscillator and $V(x) = \frac{1}{2}kx^2$. For regions of x where $|E| > \frac{1}{2}kx^2$ we might anticipate oscillatory solutions, whilst outside this region ($|E| < \frac{1}{2}kx^2$) spatially damped solutions might be expected. Indeed, at large x such that $\frac{1}{2}kx^2 \gg E$ we have the approximate equation

$$\frac{d^2\psi}{dx^2} - \frac{8\pi^2 m}{h^2}\frac{kx^2}{2}\psi \sim 0 \tag{2.8.5}$$

which has the approximate solution

$$\psi(x) \sim A \exp\left(-\frac{\beta x^2}{2}\right) + B \exp\left(\frac{\beta x^2}{2}\right) \tag{2.8.6}$$

where $\beta = \sqrt{(mk)}/\hbar$.

Since we are interested only in finite bounded wavefunctions, it follows that $B = 0$, $\psi(x)$ otherwise increasing indefinitely with x. We therefore have the approximate solution in the limit of large x

$$\psi(x) \underset{|x|\to\infty}{\to} A \exp\left(-\frac{\beta x^2}{2}\right) \tag{2.8.7}$$

We have so far dealt only with potential functions which are not continuous functions of position, and so, as we might expect, the eigenfunctions satisfying the Schrödinger equation (2.8.4) will be somewhat different from the simple trigonometric solutions obtained earlier. Since we have the asymptotic solution (2.8.7), we might suppose that solutions to equation (2.8.4) will have the form

$$\psi(x) = H(x)\,e^{-\beta x^2/2} \tag{2.8.8}$$

where $H(x)$ is a function to be determined, and our only initial requirement is that this function must allow the asymptotic solution (2.8.7) to develop in regions for which $E < \frac{1}{2}kx^2$. For small x we see from equation (2.8.8) that the functional form of $H(x)$ dominates, and all we can say so far is that it is almost certainly oscillatory.

It is convenient to make a change of variables at this stage, and we set:

$$\alpha = 2mE/\hbar^2, \quad \beta = \sqrt{(mk)}/\hbar, \quad \xi = \sqrt{\beta}x$$

The Schrödinger equation (2.8.4) now reads

$$\frac{d^2H}{d\xi^2} - 2\xi\frac{dH}{d\xi} + \left[\frac{\alpha}{\beta} - 1\right]H = 0 \tag{2.8.9}$$

where we have used the transformed wavefunction $\psi(\xi) = H(\xi)\,e^{-\xi^2/2}$. In the constant potential case many of the trigonometric solutions could be rejected on the grounds that they were not well-behaved. In the present case it is found that well-behaved solutions may only be obtained for (see Appendix IV)

$$\frac{\alpha}{\beta} = 2n + 1 \tag{2.8.10}$$

that is, for

$$E = \hbar\sqrt{(k/m)}(n + \tfrac{1}{2})$$

$$= h\nu(n + \tfrac{1}{2}), \quad \nu = \frac{1}{2\pi}\sqrt{\frac{k}{m}} \tag{2.8.11}$$

where ν is the oscillator frequency. The relation (2.8.10) is of course a statement of the energy quantization of the harmonic oscillator. We notice that the levels are separated by integral amounts $h\nu$, except for the ground state level ($n = 0$) which appears at $h\nu/2$. Indeed all the energy levels are of half-integral order, which seems at variance with our ideas of energy quantization. These *zero point effects* are, as we shall see, a direct consequence of the *uncertainty relations*.

Substitution of equation (2.8.10) in equation (2.8.9) yields

$$\frac{d^2 H_n}{d\xi^2} - 2\xi\frac{dH_n}{d\xi} + 2nH_n = 0 \tag{2.8.12}$$

which is *Hermite's equation*, and its solution is given in Appendix IV. As a solution to equation (2.8.12) we obtain

$$H_n(\xi) = k\,e^{\xi^2}\frac{d^n}{d\xi^n}(e^{-\xi^2}) \tag{2.8.13}$$

where k is an arbitrary constant. Choosing $k = (-1)^n$ we obtain the following forms of solution:

$$\left.\begin{aligned}
H_0 &= 1 \\
H_1 &= 2\xi \\
H_2 &= 4\xi^2 - 2 \\
H_3 &= 8\xi^3 - 12\xi \\
H_4 &= 16\xi^4 - 48\xi^2 + 12 \\
H_5 &= 32\xi^5 - 160\xi^3 + 120\xi
\end{aligned}\right\} \tag{2.8.14}$$

We see that the solutions are of polynomial form and that $H_n(\xi)$, the *Hermite polynomial of degree n*, contains terms up to order ξ^n. The final eigenfunctions of equation (2.8.4) are therefore

$$\psi_n(x) = H_n(\sqrt{\beta}x)\,e^{-\beta x^2/2} \tag{2.8.15}$$

and the first few eigenfunctions are shown in Figure 2.8.1. The probability amplitudes, $\psi(x)$, are quite similar to the solutions obtained in the cases of the infinite and finite rectangular wells, where the width of the well a now varies as kx^2. This we might have expected. The probability distribution $|\psi(x)|^2$ is particularly interesting in that it differs markedly from the classical distribution, particularly at low energies. In Figure 2.8.2 the probability distributions $|\psi_n|^2$ for $n = 0, 2, 5$ and 20 are compared with the classical distributions, shown by

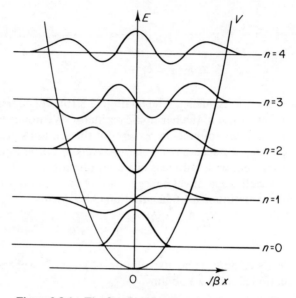

Figure 2.8.1 The first five eigenfunctions for the linear harmonic oscillator. Notice how the even ($n = 0, 2, 4, \ldots$) states are symmetric, whilst the odd states are antisymmetric (cf. Figure 2.7.6). Quantum mechanically there exists a finite probability of finding the particle beyond the *classical turning points* defined by the classical amplitude of vibration in the parabolic potential $V(x)$

the broken line. The zero-point motion corresponding to $n = 0$ is almost the exact opposite of the classical distribution. However, for very highly excited states, or for high particle energies we see that the classical distribution is regained, another example of the Bohr correspondence principle. We notice again the phenomenon of quantum tunnelling. There is evidently a finite probability of finding the particle beyond the limits of the classical amplitude of oscillation given approximately by the relation

$$|\psi|^2 \sim \exp\left[-\frac{\sqrt{(mk)}}{\hbar}x^2\right] \tag{2.8.16}$$

(cf. equation (2.8.7)) according to which the tunnelling is sensitively dependent upon the mass of the particle.

Suppose the linear harmonic oscillator represented the quantized modes of vibration of an atom in a one-dimensional lattice. What would happen as the temperature or the total energy of the system was reduced? The oscillator would eventually reach its ground state, and at absolute zero the kinetic energy and linear momentum would presumably be zero. Moreover the particle must then be located at the potential minimum, $x = 0$. We evidently have a precise and simultaneous knowledge of both the momentum and position of the particle.

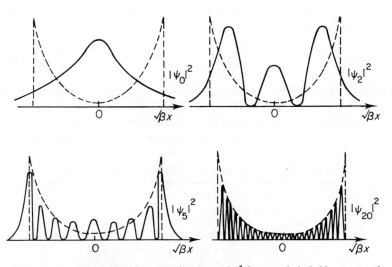

Figure 2.8.2 The probability distributions $|\psi_n|^2$ for $n = 0, 2, 5, 20$ compared with the classical distributions. There is a finite probability of finding the particle in classically forbidden regions defined by the classical amplitude of vibration. The ground state ($n = 0$) distribution is almost the exact opposite of the classical probability distribution (broken curve), although the classical limit is regained for very highly excited ($n \to \infty$) states (cf. Figure 2.7.3). This represents an example of the Bohr correspondence principle

This is clearly inconsistent with the Heisenberg uncertainty principle, according to which the intrinsic spatially indeterminate nature of the wavefunction and its reciprocal relation to the momentum (via the de Broglie relation) requires that $\Delta p \Delta x \geqslant \hbar$.

We may quite easily show that the zero point oscillation of energy $E_0 = h\nu/2$ for the linear harmonic oscillator is entirely consistent with the uncertainty principle. The momentum of the particle in the ground state is $p = \pm\sqrt{(2mE_0)}$ (the average momentum, of course, being zero) and the uncertainty in our knowledge of the momentum is therefore $\Delta p \sim 2\sqrt{(2mE_0)}$. The zero point amplitude of vibration is approximately given as $2\sqrt{(E_0/k)}$, so that we have

$$\Delta p \Delta x \sim 4E_0 \sqrt{\left(\frac{2m}{k}\right)}$$

$$\sim \frac{2h\nu}{2\pi\nu} \geqslant \hbar$$

since $\nu = (2\pi)^{-1}\sqrt{(k/m)}$. Thus for any constrained or confined motion we must anticipate the existence of zero point effects, even for a rectangular well. These effects are most marked for particles of low mass, and, of course, at low temperatures when the system is essentially in its ground state. A very useful measure of the susceptibility of a system to zero point effects is provided by the *de Boer parameter* Λ, defined as

$$\Lambda = \frac{h}{d\sqrt{(mkT)}} = \frac{\lambda}{d} \tag{2.8.17}$$

where d is the mean atomic spacing of the system and $\sqrt{(mkT)}$ is the momentum associated with particles of mean energy $3kT/2$. The de Boer parameter represents the ratio of the thermal de Broglie wavelength λ to the mean atomic spacing, where the energy has been expressed in units of kT. In Table 2.8.1 we

Table 2.8.1. De Boer parameters for the liquid inert gases

	At. wt.	Λ
He	4·003	0·424
Ne	20·138	0·0939
Ar	39·944	0·0294
Kr	83·7	0·0161
Xe	131·3	0·0103

compare the de Boer parameters for a number of liquid inert gases at their triple points, and we see that only for helium are zero point effects likely to be important. Indeed, the zero point oscillations in liquid helium are sufficient to prevent the condensation of the liquid phase into a stable solid even at absolute zero. Only under the application of an external pressure of 25 atmospheres in the case of He^4, and 28 atmospheres for He^3, does the liquid eventually condense into a stable crystalline solid. The extra three atmospheres in the case of He^3 may be traced to the additional statistical repulsion developed in an assembly of fermions as a consequence of exclusion conditions placed on the atomic wavefunctions. The pressure–temperature diagrams for these two systems are shown in Figure 2.8.3.

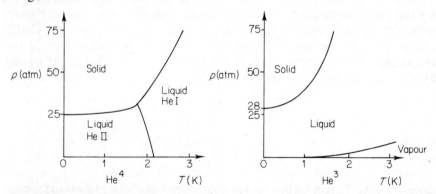

Figure 2.8.3 The p–T diagrams for He^3 and He^4. In both systems the zero point motions are such that a stable solid phase develops at absolute zero only under an applied pressure of ~ 25 atm for the boson (He^4) system, and ~ 28 atm for the fermion (He^3) system. The extra three atmospheres in the case of He^3 is required to account for the additional 'statistical repulsion' developed between fermions

The case of two identical coupled harmonic oscillators is interesting in that it represents the only separable case for interacting particles. It also demonstrates the removal of degeneracy in the two-body system as the coupling is increased.

Clearly, for two *independent* (uncoupled) linear harmonic oscillators the total energy of the system will be the sum of the individual energies:

$$E_n = \hbar\omega(n_1 + \tfrac{1}{2}) + \hbar\omega(n_2 + \tfrac{1}{2}) \qquad (2.8.18)$$

where n_1 and n_2 are integers chosen such that $n = n_1 + n_2$. It is evident also that the system is $(n + 1)$-fold degenerate, including the ground state, n_1 or $n_2 = 0$. If the location of the two particles on the x-axis is at x_1 and x_2 and the coupling is such that the force between the two oscillators is $c(x_1 - x_2)$ where c is a coupling constant, then the Schrödinger equation becomes (cf. equation (2.8.4))

$$\frac{\partial^2 \psi}{\partial x_1^2} + \frac{\partial^2 \psi}{\partial x_2^2} + \left(\frac{2mE}{\hbar^2} - \beta^2 x_1^2 - \beta^2 x_2^2 - \gamma^2(x_1 - x_2)^2 \right)\psi \qquad (2.8.19)$$

where $\beta = m\omega/\hbar = \sqrt{(mk)}/\hbar, \gamma = \sqrt{(cm)}/\hbar$. This equation may be separated if we write

$$\xi = \frac{1}{\sqrt{2}}(x_1 + x_2); \qquad \eta = \frac{1}{\sqrt{2}}(x_1 - x_2)$$

giving

$$\frac{\partial^2 \psi}{\partial \xi^2} + \frac{\partial^2 \psi}{\partial \eta^2} + \left(\frac{2mE}{\hbar^2} - \beta^2 \xi^2 - \beta_+^2 \eta^2 \right)\psi = 0 \qquad (2.8.20)$$

where $\beta_+^2 = \beta^2 + \gamma^2 = (c + k)m/\hbar = (m\omega_+/h)^2$ and where $\omega_+^2 = \omega^2 + c/m$. Not surprisingly the eigenfunctions turn out as simple (modified) products of Hermite functions as discussed in the single-particle case (equation (2.8.15)):

$$\psi_n(\xi, \eta) = A(n_1, n_2)\, e^{-\frac{1}{2}\beta \xi^2}\, e^{-\frac{1}{2}\beta_+ \eta^2} H_{n_1}(\xi\sqrt{\beta}) H_{n_2}(\eta\sqrt{\beta_+}) \qquad (2.8.21)$$

where $A(n_1, n_2)$ is a normalization constant, and has the associated energies (where $_c$ denotes coupling)

$$_cE_n = \hbar\omega(n_1 + \tfrac{1}{2}) + \hbar\omega_+(n_2 + \tfrac{1}{2}); \qquad n = n_1 + n_2 \qquad (2.8.22)$$

which are no longer degenerate. It should be understood that *both* quantum numbers n_1 and n_2 apply to *both* particles and not to the individual oscillators. If $\omega_+ \rightarrow \omega$, i.e. if $c \rightarrow 0$ then the degeneracy would recur and equation (2.8.22) would revert to equation (2.8.18). As it is, the coupling *raises* the energy level and splits the nth degenerate state into $(n + 1)$ levels; the first few are shown explicitly in Figure 2.8.4 as the coupling is progressively increased. It should be noted that $\omega_+ > \omega$ and so for the nth state the lowest level will be $(n + \tfrac{1}{2})\hbar\omega + \tfrac{1}{2}\hbar\omega_+$, whilst the highest will be $\tfrac{1}{2}\hbar\omega + (n + \tfrac{1}{2})\hbar\omega_+$. We should also observe that the symmetry of the function $H_{n_2}(\eta\sqrt{\beta_+})$ changes with respect to exchange

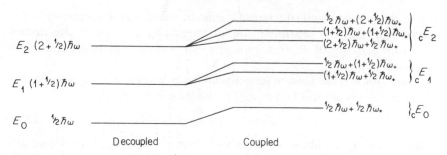

$\frac{1}{2}\hbar\omega + (2+\frac{1}{2})\hbar\omega_+$
$(1+\frac{1}{2})\hbar\omega + (1+\frac{1}{2})\hbar\omega_+$ $\left.\right\}_c E_2$
$(2+\frac{1}{2})\hbar\omega + \frac{1}{2}\hbar\omega_+$

$\frac{1}{2}\hbar\omega + (1+\frac{1}{2})\hbar\omega_+$
$(1+\frac{1}{2})\hbar\omega + \frac{1}{2}\hbar\omega_+$ $\left.\right\}_c E_1$

$\frac{1}{2}\hbar\omega + \frac{1}{2}\hbar\omega_+$ $\left.\right\}_c E_0$

E_2 $(2+\frac{1}{2})\hbar\omega$

E_1 $(1+\frac{1}{2})\hbar\omega$

E_0 $\frac{1}{2}\hbar\omega$

Decoupled   Coupled

Figure 2.8.4 Splitting of degenerate eigenstates for a system of two coupled linear harmonic oscillators

of the particle coordinates (η changing sign) according as n_2 is odd or even—see equation (2.8.14). The $H_{n_1}(\xi\sqrt{\beta})$ function is unaffected by the exchange. On this basis the states are said to have *odd* or *even symmetry* as n_2 is odd or even, respectively.

2.9 The Periodic Potential

A useful model potential for a conduction electron in a crystalline lattice is the one-dimensional sinusoidal potential

$$V(x) = \left\{ V_0 - V_1 \cos\left(\frac{2\pi x}{a}\right) \right\} \tag{2.9.1}$$

where V_0 is a negative constant, V_1 is the amplitude of the potential function and a is the lattice spacing. This function gives a reasonable representation of the actual periodic potential in the crystalline lattice for conduction electrons: the two are compared in Figure 2.9.1. Another example of the periodic potential

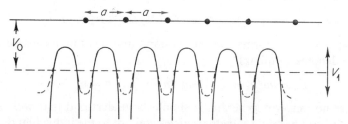

Figure 2.9.1 One-dimensional lattice potential (full curve), and its approximate representation by the sinusoidal function $V(x) = V_0 - V_1 \cos(2\pi x/a)$, where V_0 is the negative mean 'trough' potential

occurs for the rotation of one methyl group relative to the other about the C—C bond in the ethane molecule (Figure 2.9.2). Again the potential is of the form (2.9.1) although now x represents the relative angular displacement of the two

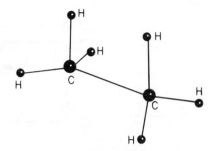

Figure 2.9.2 The ethane molecule. The two methyl groups CH_3 are free to rotate about the $C-C$ bond, although there will obviously be relative orientations of lower potential energy: the methyl groups may be imagined as 'clicking' into the minimum energy position

methyl groups and a^{-1} is a constant numerically equal to three, accounting for the 120° rotational symmetry.

When the Schrödinger equation is solved in a region where the potential is a periodic function of position the permitted eigenvalues appear in *bands* of values with forbidden energy gaps between them. Solutions in these forbidden regions represent spatially damped or attenuated wavefunctions and as such do not contribute to the electronic conduction process. The development of a band theory of solids has been fundamental in the description of the properties of conductors, insulators and semiconductors. There is a close similarity here between the optical and acoustic modes of vibration and their associated dispersion relations for a one-dimensional lattice discussed in Section 2.5.

First of all we can consider the development of energy bands in an assembled lattice of identical atoms from a purely physical standpoint. The electronic energy levels of two widely separated atoms will of course be identical and therefore degenerate, but as the atoms approach one another coupling or perturbation of the orbitals will occur and hybrid wavefunctions will develop. At the same time the degeneracy of the eigenfunctions in the separated configuration will be removed and the single degenerate state for two isolated atoms will split into two states as shown in Figure 2.9.3(a). This splitting of the degenerate levels as the system is condensed may be understood as follows. We have seen (Section 2.3) that the hybrid wavefunctions which develop may be constructed from the 'normal mode' solutions of the individual atoms. Thus if ψ_1 is taken to represent the atomic orbital of electron 1 of energy E_0 on the first atom, and ψ_2 represents the atomic orbital of electron 2 on the second atom, again of energy E_0, then we may construct the unnormalized hybrid wavefunctions for the two electrons (linear combination of atomic orbitals; LCAO)

$$\psi = \psi_1 \pm k\psi_2 \tag{2.9.2}$$

58

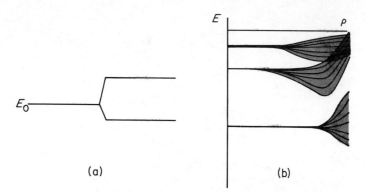

Figure 2.9.3 The energy level E_0 in two identical atoms is, for the two-atom system, effectively doubly degenerate. As the two atoms interact the degenerate state splits as shown in (a). As a system of particles is brought together, the initially multiply degenerate state broadens into a *band* of energy levels (b). The tightly bound inner states are relatively unaffected, except at very high densities. The bands may actually overlap, and indeed frequently do

where k is a coupling or hybridizing coefficient. The energy associated with such a distribution is of course

$$E = \frac{\int \psi H \psi \, d\tau}{\int \psi \psi \, d\tau} \tag{2.9.3}$$

where H is the Hamiltonian for the system, the details of which do not concern us here. The integrals range over all space, and for well-behaved wavefunctions ψ which vanish at large distances we should obtain a finite value for the energy E of the coupled system. Substituting equation (2.9.2) in equation (2.9.3) we obtain

$$\begin{aligned} E &= \frac{\int \psi_1 H \psi_1 \, d\tau + \int \psi_2 H \psi_2 \, d\tau \pm 2k \int \psi_1 H \psi_2 \, d\tau}{\int \psi_1^2 \, d\tau + \int \psi_2^2 \, d\tau \pm 2k \int \psi_1 \psi_2 \, d\tau} \\ &= \frac{E_0 \pm J}{1 \pm S} \end{aligned} \tag{2.9.4}$$

where $J = k \int \psi_1 H \psi_2 \, d\tau$ is termed the *exchange integral*, and represents the energy associated with the interaction of electron 1 with nucleus 2, and *vice versa*. Clearly this integral depends sensitively upon the density or degree of overlap of the atomic wavefunctions. S is the *overlap integral* and is a measure of the overlap of the atomic orbitals. Thus, at large atomic separations $J = S = 0$ and from equation (2.9.4) we regain the doubly degenerate level of energy E_0. As the density increases so do J and S and the energy level splits as shown in Figure 2.9.3(a). (Note that the same signs are taken together in equation (2.9.4), i.e. $+/+$ and $-/-$). This argument may be simply extended to a system of N atoms when the N-fold degenerate energy levels of the separated system split to form energy bands as shown in Figure 2.9.3(b). Clearly the outermost energy levels will overlap most extensively as the lattice condenses, whilst the inner shells hybridize to a correspondingly smaller extent. These latter levels are generally filled, and do not participate in the electronic conduction process. We see both from equation (2.9.4) and Figure 2.9.3(b) that the spread of the energy bands will depend to a great extent upon the density and chemical constitution of the assembly, and in some cases the bands will actually overlap, leading to more complex conduction processes.

The one-dimensional Schrödinger equation for an electron in the periodic potential (2.9.1) is

$$\frac{d^2\psi}{dx^2} + \frac{8\pi^2 m}{h^2}\left(E - V_0 + V_1 \cos\frac{2\pi x}{a}\right)\psi = 0 \qquad (2.9.5)$$

If we make the following substitutions

$$\xi = \frac{\pi x}{a}; \quad \alpha = \frac{8ma^2}{h^2}(E - V_0); \quad \beta = \frac{8ma^2}{h^2}V_1$$

then equation (2.9.5) reduces to

$$\frac{d^2\psi}{d\xi^2} + (\alpha + \beta \cos 2\xi)\psi = 0 \qquad (2.9.6)$$

which is *Mathieu's equation*, and its solution is given in Appendix V. (The Mathieu solutions and their applications to problems in an elliptical coordinate frame are discussed at some length in Section 3.10.) We see that for $\beta = 0$ (corresponding to $V_1 = 0$) representing a region of constant potential $-V_0$, we regain the simple trigonometric solutions to equation (2.9.6). The regions of solution to the Mathieu equation are given in Figure 2.9.4 as functions of the parameters α and β. Only solutions to the Mathieu equation corresponding to values of α and β which lie within the shaded regions of Figure 2.9.4 represent unattenuated wavefunctions: outside these regions the solutions are spatially damped and do not interest us here. Thus, for a given value of $\beta = 8ma^2V_1/h^2$, that is for a given value of the lattice separation and amplitude of the periodic potential, unattenuated solutions are obtained for regions of α given by the intersection of the vertical line with the shaded areas. Only certain *bands* of energy are therefore permitted since only certain values of the function $\alpha = 8(E - V_0)ma^2/h^2$ correspond to unattenuated solutions, for a given value of β. We therefore conclude that it is possible to find only certain ranges of energy which are permissible in the solution of the Schrödinger equation for a periodic potential.

It is interesting to observe how the energy bands develop as a function of the parameter $\beta = 8ma^2V_1/h^2$. If we regard V_1 as fixed, and expand the lattice, that is, increase the lattice spacing a (shown by the broken line in Figure 2.9.4), we see that the bands degenerate with increasing a. Ultimately, at low densities, we regain the discrete, highly degenerate levels shown in Figure 2.9.3(b). Moreover

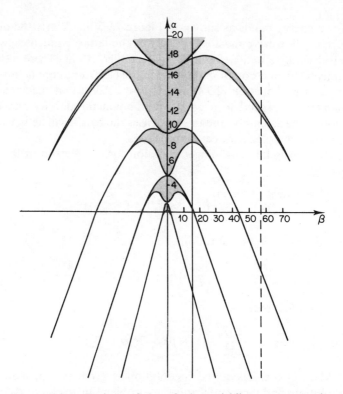

Figure 2.9.4 Regions of the $\alpha\beta$ plane yielding unattenuated solutions to Mathieu's equation (shaded areas) for a periodic potential. Solutions outside these regions correspond to 'Tamm states', and represent spatially damped solutions. For a given value of the parameter β unattenuated solutions are obtained for those values of α given by the intersection of the vertical broken line with the shaded areas. Large values of the parameter β evidently correspond to highly degenerate solutions: the precise physical interpretation of α and β depends upon the details of the problem

the width of the bands increases with increasing energy as we anticipated from the purely physical discussion above. Suppose we now consider the case for $V_1 \to 0$, representing a diminishing amplitude in the oscillation of the periodic potential, until for $\beta = 0$ the potential assumes the constant negative value of V_0. Then we see from Figure 2.9.4 that the energy bands have spread to form a *continuum* of states: any subsequent quantization of the energy levels can only arise as a consequence of boundary conditions imposed by the surfaces of the metal.

The spatial form of the wavefunction in the periodic potential is shown schematically in Figure 2.9.5. The probability amplitudes for the conduction electrons maximize in the vicinity of the ionic cores as we might expect from the form of the potential function. Bloch, as a development of le Floquet's theorem,

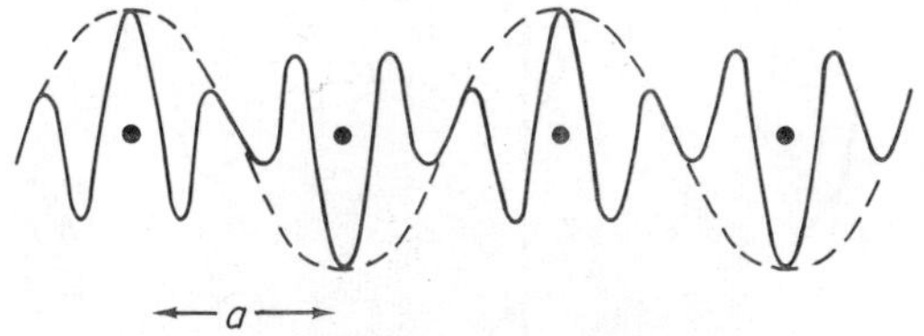

Figure 2.9.5 Bloch modulation of the electronic
wavefunction by a periodic potential

has proposed that the wavefunction should have a periodic envelope related to
the periodicity of the lattice. We therefore consider the electronic wavefunction
in the conduction band to consist of the product

$$\psi_k(x) = u_k(x)\, e^{ikx} \tag{2.9.7}$$

where x is the position vector and $u_k(x)$ has the translational periodicity of the
crystal. As we observed for the case of propagation of mechanical lattice waves
(Section 2.5) as we pass through the Bragg standing wave condition with
increasing k, so the energy flux through the lattice will diminish and increase as
$k = \pm n\pi/a$. Thus in spite of an applied electric field, the classical relationship
between force and energy $dE/dt = F_x v_x$ will not apply. We are not setting
Newton's second law of motion aside; the point is that this neglects the electron–
lattice interaction arising from the scattering of the electronic wavefunctions
from the potential variations within the lattice. It is hardly suprising that the
resulting equation of motion is not completely described in terms of the
externally applied field. Any deviation from the simple equation of motion may,
however, be interpreted in terms of an *effective mass, m^**.
    Thus,

$$\frac{dE}{dk} = \frac{1}{2m^*}\frac{d}{dk}(p^2) = \frac{\hbar^2 k}{m^*} \tag{2.9.8}$$

from the de Broglie relation $p = h/\lambda$ and $k = 2\pi/\lambda$. Therefore

$$m^* = \frac{\hbar^2}{d^2E/dk^2} \tag{2.9.9}$$

We observe from equation (2.9.8) that $dE/dk = hv_e/4\pi^2$, where v_e is the velocity
of propagation of the electron within the crystal. It follows that the gradient
dE/dk is zero at the Bragg condition, and this is evident from Figure 2.9.6.
Remembering that only certain energy bands are permitted, and that the
intervening gaps are forbidden, we obtain an $E(k)$ dispersion relation for an
electron in a one-dimensional periodic potential of the schematic form shown
in Figure 2.9.6.
    We might well enquire just what is going on physically at these particular
values of the electron momentum or wavenumber corresponding to the discontinuities in the $E(k)$ curve. We know that these particular wavenumbers

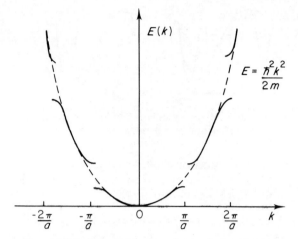

Figure 2.9.6 Dispersion curve for an electron in a one-dimensional periodic potential. The classical parabolic $E(k)$ relationship breaks up into a series of 'allowed' zones separated by 'forbidden' energy gaps corresponding to spatially damped non-propagating Tamm states. For a lattice of period a the discontinuities in the $E(k)$ curve occur at integral multiples of $k = \pm\pi/a$. In the case of a three-dimensional system we would obtain a *dispersion surface* whose precise form would depend upon the lattice parameters a, b, c

satisfy the Bragg condition for standing waves in a one-dimensional periodic potential. Now, we may easily construct standing waves for a *free* electron by combining two travelling waves—one moving to the left and one moving to the right, thus:

$$\psi_\pm = \frac{1}{\sqrt{2}}\{\exp(ikx) \pm \exp(-ikx)\} = \begin{pmatrix} \sqrt{2}\cos kx \\ \sqrt{2}i\sin kx \end{pmatrix}$$

Moreover, both these waves have the same energy. For such standing waves in a periodic potential, however, these two linear combinations will have *different* energies since one localizes the electrons on the potential minima, and the other on the potential maxima. If we actually calculate the average potential energy in each state we have

$$V_\pm = \frac{1}{L}\int_0^L |\psi_\pm|^2 V(x)\,\mathrm{d}x$$

$$= \int_0^L \begin{pmatrix} 2\cos^2 kx \\ 2\sin^2 kx \end{pmatrix} V(x)\,\mathrm{d}x = \pm|V_g|$$

where $L \gg k^{-1}$ is the length of the crystal and $g = 2\pi/a$ is the fundamental wavenumber of the lattice potential. If we now consider the sum of the kinetic and potential energy components at $k = \pi/a$, for example, then of course the kinetic energy will be identical to the free electron value: $\hbar^2 k^2/2m$. Thus the total energy of these two states is

$$E_{k=\pi/a} = \frac{\hbar^2 k^2}{2m} \pm |V_g|$$

Evidently the total energy is depressed below the free electron value for the distribution corresponding to the location of the probability maxima on the potential minima, and vice versa—both these components being present to the same extent at the Bragg condition.

We see from equation (2.9.9) that the effective electron mass is inversely proportional to the curvature d^2E/dk^2, and for an electron near the top of one of the allowed energy bands this is seen to be negative. The physical significance of a particle with a negative effective mass is that it behaves as if it had a positive charge under the action of electric and magnetic fields. Electrons that reside above the inflection point of $E(k)$ evidently behave like positive charges, and as such offer an explanation of the anomalous Hall effect: metals and semiconductors having a positive Hall coefficient have nearly filled bands. Heisenberg has shown that it is possible to replace the wave equation for the electrons in a nearly filled band by an approximate equation for the holes in the band, to which these electrons of negative effective mass are formally equivalent. In Table 2.9.1 we list some ratios of the effective to the true electron mass.

Table 2.9.1. Effective electron masses

Element	m_e^*/m_e
Li	1·40
Na	0·98
K	0·94
Rb	0·87
Cs	0·83
Cu	1·47
Al	2·30

The region of the dispersion curve in the interval $k = \pm\pi/a$ is termed the *first Brillouin zone*, the interval $\pm\pi/a \leqslant k \leqslant \pm 2\pi/a$ the *second Brillouin zone*, and so on. The parabolic shape of the dispersion curve appropriate to a particle moving in a region of constant potential is preserved, except that forbidden energy gaps develop. We also see that a modification of the parabolic form develops in regions where the wavenumber of the electron moving in the one-dimensional lattice approaches some integral multiple of the wavenumber of the periodic potential.

With regard to our interest in conduction properties, only the uppermost partially filled energy band is of interest. The lower-lying filled bands corresponding to the filled degenerate atomic levels of the expanded system do not participate directly in the conduction process. Confining our interest to the uppermost energy band we understand that for insulators this band of doubly degenerate (spin up, spin down) electronic states is completely filled. There is no possibility of electrons acquiring additional energy under the action of an applied electric field since the adjacent states are forbidden. Moreover, the size of the energy gap (~ 6 eV in diamond) is sufficiently large that at room temperatures ($\sim 1/40$ eV) there is no possibility of thermal excitation into a higher band. Metals, on the other hand, are characterized by an upper band in which the levels are partially filled to an energy E_F, the Fermi energy. It is clear that in this case conduction may take place easily in the vacant states within the band (Figure 2.9.7).

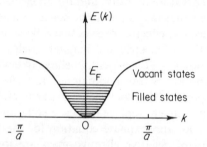

Figure 2.9.7 The uppermost energy band may only be partially filled, and this characterizes the metals. Each of the filled levels in the band are doubly occupied by electrons of opposed spin at absolute zero, up to a maximum energy level E_F, the Fermi energy. Electrons in the vicinity of the Fermi surface are able to acquire energy under the action of an applied electric field, populating the vacant states, and are consequently able to participate in the conduction process

For a three-dimensional lattice the dispersion curve will develop into constant energy surfaces in k-space. The Fermi surface for a simple lattice is shown in Figure 2.9.8(a), and for a more complicated lattice in Figure 2.9.8(b): clearly the shape of the Fermi surface is governed by the reciprocal lattice structure.

We may now return to consider the quantization of the rotational energy of the methyl group in the ethane molecule (Figure 2.9.2). For this molecule the potential function is given as

$$V(\theta) = -V_1 \cos n\theta \qquad (2.9.10)$$

where θ is the relative angular displacement of the two methyl groups; $n = 3$ and $V_1 \sim 1\cdot5$ kcal/mole for ethane. The numerical parameters n and V_1 in equation (2.9.10) will vary from molecule to molecule of course. If we now make

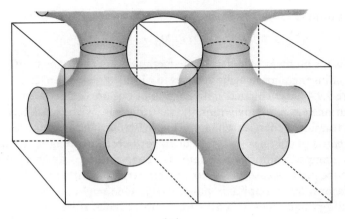

(a)

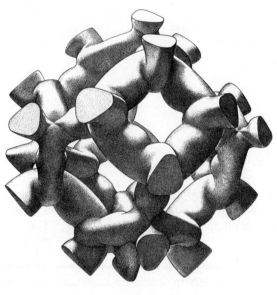

(b)

Figure 2.9.8 The Fermi surface is an imaginary surface in momentum space below which the electronic energy levels are completely filled, and above which they are completely vacant at absolute zero. At temperatures above zero the sharp Fermi surfaces shown here are somewhat blurred by thermal excitation of electrons just below the Fermi surface to the vacant states immediately above. (a) shows a very simple Fermi surface, whilst in (b) we show the 'third zone' of a very complicated Fermi surface —that of the metal lead

the substitutions

$$\xi = 3\theta/2; \quad \alpha = 2IE/(3\hbar)^2; \quad \beta = IV_1/2(3\hbar)^2$$

where I is the moment of inertia of the methyl group about the C—C axis, then the Schrödinger equation again reduces to Mathieu's equation (2.9.6). Referring to Figure 2.9.4, unattenuated stationary solutions are obtained only for values of the parameters α, β lying within the shaded areas. We immediately conclude that the rotational energies which the methyl groups may assume fall into well defined *bands* of energies with forbidden gaps corresponding to attenuated or spatially damped solutions to Mathieu's equation. It is again instructive to consider the limiting case as β becomes very large, due either to a high moment of inertia or a large amplitude potential V_1. One might visualize this case as representing high energy 'click-stops', three click-stops representing one complete angular rotation of the methyl group for this molecule. Clearly the methyl group will most likely be found in the angular configurations $\theta \sim 0$, $2\pi/n, 4\pi/n, \ldots, 2\pi$, where $n = 3$ in this case. If we therefore make the approximation $\cos 2\xi \sim 1 - 2\xi^2$, then the Mathieu equation becomes

$$\frac{d^2\psi}{d\xi^2} + (4\alpha - 16\beta + 32\beta\xi^2)\psi = 0 \tag{2.9.11}$$

and if we further set

$$\Xi = 4\sqrt{(32\beta)}; \quad \gamma = (\alpha + 4\beta)/\sqrt{(2\beta)} \tag{2.9.12}$$

then equation (2.9.11) becomes

$$\frac{d^2\psi}{d\Xi^2} + (\gamma - \Xi^2)\psi = 0 \tag{2.9.13}$$

and this is of the same form as for the linear harmonic oscillator, equation (2.8.4). Again, the solutions are only well behaved if $\gamma = (2K + 1)$ where K is an integer. Under these circumstances, from equation (2.9.12)

$$\alpha = 4\beta + (2K + 1)\sqrt{(2\beta)}$$

and the associated discrete energy levels are, from the former set of substitutions

$$E = V_1 + 3\hbar\sqrt{(V_1/I)}(K + \tfrac{1}{2}) \tag{2.9.14}$$

There is seen to be a zero point rotational energy corresponding to $K = 0$. These discrete levels appear in Figure 2.9.4 at large β. As $\beta \to 0$, that is, as either or both the moment of inertia of the rotating group or the amplitude of the potential function tends to zero, the Mathieu equation simplifies to give in the limit:

$$\frac{d^2\psi}{d\theta^2} + 9\alpha\psi = 0 \tag{2.9.15}$$

in the original variables. This corresponds to very weak 'click-stop' positions of the rotating group. Well-behaved solutions to equation (2.9.15) must, of course, obey the condition:

$$\psi(\theta) = \psi\left(\theta + \frac{2\pi}{3}\right) \tag{2.9.16}$$

Consequently acceptable solutions take the form

$$\psi(\theta) = \exp \pm i3\sqrt{\alpha}\theta$$

where $3\sqrt{a} = M = 0, 1, 2, \ldots$. The rotational energy levels are therefore

$$E = (M\hbar)^2/2I \tag{2.9.17}$$

A free rotator has eigenvalues $E = (m\hbar)^2/2I$ where $M = 3m$, but we see here that the symmetry of the problem allows only every third eigenstate to develop. It is interesting to observe that the condition (2.9.16) effectively quantizes the rotational energy spectrum, and the continuum of states shown in Figure 2.9.4 for $\beta = 0$ does not develop.

2.10 The Diffusion Equation

In Section 1.3 we developed the general diffusion equation

$$D\nabla^2\psi(\mathbf{r}, t) = \frac{\partial\psi(\mathbf{r}, t)}{\partial t} + a(\mathbf{r}, t) - c(\mathbf{r}, t)$$

where D is the diffusion coefficient in an isotropic homogeneous medium and c and a are creation and annihilation source–sink terms. This equation adopts a particularly simple form when there is no source–sink distribution, and $\partial\psi/\partial t = 0$, reducing then to Laplace's equation.

We consider first of all one-dimensional solutions to the diffusion equation in a region where there is no source–sink distribution. Under these circumstances.

$$\frac{\partial^2\psi(x, t)}{\partial x^2} = \frac{1}{D}\frac{\partial\psi(x, t)}{\partial t} \tag{2.10.1}$$

where ψ might represent some scalar field such as the distribution of concentration or temperature in the case of a thermal conduction field. The diffusion coefficient is then termed the *thermal diffusivity* (= thermal conductivity/ (specific heat × density)) of the medium.

Trying a separation of variables

$$\psi(x, t) = X(x)T(t)$$

we obtain the separated equations

$$\frac{d^2 X}{dx^2} + k^2 X = 0 \qquad\qquad (2.10.2a)$$

$$\frac{dT}{dt} + Dk^2 = 0 \qquad\qquad (2.10.2b)$$

where k^2 is the separation constant. The spatial equation in $X(x)$ is of Helmholtz form, and we might anticipate oscillatory solutions. The equation describing the time evolution appears from (2.10.2) to give rise to temporally damped solutions.

A fundamental distinction between the solutions of the diffusion equation and the solutions of the wave equation should be emphasized at this point. The solutions differ on account of the *irreversibility* introduced in the diffusion equation by the first time derivative. The wave equation on the other hand, containing the second time derivative, is essentially *reversible* in time as the following two examples will show.

A flexible string of length a held in the initial configuration $\psi_0(x)$ and then released executes the motion

$$\psi(x, t) = \sum_{n=1}^{\infty} A_n \sin\left(\frac{n\pi x}{a}\right) \cos\left(\frac{n\pi c t}{a}\right)$$

$$A_n = \frac{2}{a} \int_0^a \psi_0(x) \sin\left(\frac{n\pi x}{a}\right) dx$$

where c is the wave velocity. In the case of the diffusion process, a slab of material thickness a having the initial temperature distribution $\psi_0(x)$ and its opposite faces held at zero temperature will have the subsequent temperature distribution

$$\psi(x, t) = A_n \sin\left(\frac{n\pi x}{a}\right) \exp\left[-\left(\frac{n\pi}{a}\right)^2 Dt\right]$$

$$A_n = \frac{2}{a} \int_0^a \psi_0(x) \sin\left(\frac{n\pi x}{a}\right) dx$$

The spatial factors are identical—the time dependences are quite different. Indeed, in the case of the wave equation the solution does not change its form if the time is reversed, whilst in the case of diffusion the solution becomes divergent. This we can understand in terms of the second time derivative remaining invariant to the transformation $t \to -t$, which is not the case for the first derivative. As an example of the use of equations (2.10.2) we may consider the problem of the semi-infinite slab where the function $\psi(0, t)$ varies sinusoidally at the surface. Such a problem is reasonably represented by the space–time variation of the thermal field below the surface of the earth due to the sun. We could be more precise and incorporate the annual as well as the diurnal sinusoidal variation about the mean surface temperature, but we shall restrict the present discussion to the simple sinusoidal variation. We have the boundary conditions

$$\psi(0, t) = S\,e^{i\omega t} \quad (t > 0)$$
$$\psi(x, 0) = 0 \qquad (t = 0)$$
$$\psi(\infty, t) = 0 \qquad (t > 0)$$

$$(2.10.3)$$

S represents the amplitude of the surface fluctuations in the field ψ. We assume that the distribution below the surface varies sinusoidally with time since the surface value $\psi(0, t)$ varies as shown in equations (2.10.3). We therefore assume a solution of the form

$$\psi(x, t) = X(x)\,e^{i\omega t} \tag{2.10.4}$$

where $X(x)$ represents the spatial variation of the distribution. Substitution of equation (2.10.4) into the diffusion equation yields

$$\frac{d^2 X(x)}{dx^2} = \frac{i\omega}{D} X(x) \tag{2.10.5}$$

It immediately follows that solutions to equation (2.10.5) are of the form

$$X(x) = A \exp\left(\sqrt{\frac{i\omega}{D}}x\right) + B \exp\left(-\sqrt{\frac{i\omega}{D}}x\right) \tag{2.10.6}$$

where A and B are arbitrary constants. Now,

$$\sqrt{i} = \exp\left(\frac{i\pi}{4}\right) = \frac{1}{\sqrt{2}} + i\frac{1}{\sqrt{2}}$$

whereupon equation (2.10.6) may be rewritten in the form

$$X(x) = A \exp\left(\sqrt{\frac{\omega}{2D}}x\right)\exp\left(i\sqrt{\frac{\omega}{2D}}x\right) + B \exp\left(-\sqrt{\frac{\omega}{2D}}x\right)\exp\left(-i\sqrt{\frac{\omega}{2D}}x\right)$$

$$(2.10.7)$$

It is clear that for well-behaved solutions at large x, and from boundary condition (2.10.3) that $A = 0$. The first of the boundary conditions (2.10.3) enables us to write as the final distribution

$$\psi(x, t) = S \exp\left(-\sqrt{\frac{\omega}{2D}}x\right)\exp i\omega\left(t - \frac{x}{\sqrt{(2D\omega)}}\right) \tag{2.10.8}$$

First, we see that the waves propagate with a velocity $\sqrt{(2D\omega)}$. The distance $\delta = \sqrt{(2D/\omega)}$ is generally termed the skin depth, and in terms of the spatial component represents the distance within which the function $\psi(x, t)$ falls to e^{-1} of its value at the surface. This shows that the velocity of the thermal wave is greater for the diurnal than for the annual (since $\omega_d > \omega_a$), but also that the damping is greater for the former. The phase of the progressive thermal wave varies with distance as shown by the term $x/\sqrt{(2D\omega)}$ in the oscillatory factor in

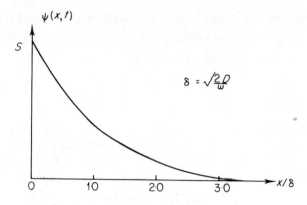

Figure 2.10.1 The envelope to the spatially damped thermal wave

equation (2.10.8). The spatial envelope of the distribution is shown in Figure 2.10.1, and the linear variation of phase with depth is shown in Figure 2.10.2. We see that the sinusoidal variation is in antiphase to the surface variation at $x/\delta = \pi = 3\cdot14$.

Another closely related problem is the solution of the diffusion equation in a semi-infinite slab for a constant surface field satisfying the following boundary conditions:

$$\left. \begin{aligned} \psi(0, t) &= S \quad (t > 0) \\ \psi(x, 0) &= 0 \quad (t = 0) \\ \psi(\infty, t) &= 0 \quad (t > 0) \end{aligned} \right\} \tag{2.10.9}$$

We have here to deal directly with the diffusion equation

$$\frac{\partial^2 \psi(x, t)}{\partial x^2} = \frac{1}{D} \frac{\partial \psi(x, t)}{\partial t} \tag{2.10.10}$$

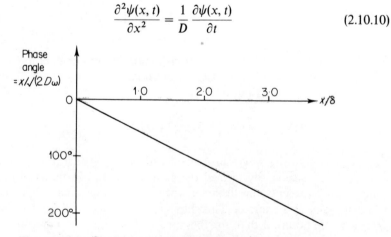

Figure 2.10.2 The linear variation of phase of the thermal wave with depth. The sinusoidal variation is in antiphase at $x = \pi$

A standard method of solution of differential equations is in terms of their Laplace transforms. We see from Appendix II that the Laplace transform of equation (2.10.10) is

$$\frac{d^2\hat{\psi}(x, p)}{dx^2} = \frac{p}{D}\hat{\psi}(x, p) - \hat{\psi}(x, 0) \tag{2.10.11}$$

where $\hat{\psi}(x, p)$ is the Laplace transform of $\psi(x, t)$. $\psi(x, 0) = 0$ from the boundary conditions (2.10.9), and consequently equation (2.10.11) has solutions of the form

$$\hat{\psi}(x, p) = A(p)\exp\sqrt{\frac{p}{D}}x + B(p)\exp-\sqrt{\frac{p}{D}}x \tag{2.10.12}$$

where $A(p)$ and $B(p)$ are arbitrary coefficients, and may be determined from the boundary conditions. These must be first transformed before they can be applied, thus:

$$\hat{\psi}(0, p) = \frac{S}{p}$$

and

$$\hat{\psi}(\infty, p) = 0$$

$$\tag{2.10.13}$$

It therefore follows that $A(p) = 0$, and $B(p) = S/p$, so that the final solution is

$$\hat{\psi}(x, p) = \frac{S}{p}\exp\left(-\frac{p}{D}x\right) \tag{2.10.14}$$

It now only remains to invert the Laplace transform $\psi(x, p)$ to regain $\psi(x, t)$ which we may do in terms of the complementary error function erfc, thus (see Appendix II, Table II.1):

$$\psi(x, t) = S\,\text{erfc}\,\frac{x}{2\sqrt{(Dt)}} \tag{2.10.15}$$

where erfc is related to the error integral erf as follows

$$\text{erfc } y = 1 - \text{erf } y$$

$$= 1 - \frac{2}{\sqrt{\pi}}\int_0^y e^{-\alpha^2}\,d\alpha \tag{2.10.16}$$

This particular form of integrand arises in the theory of random errors and describes their Gaussian distribution about the mean. The normalized integral over the range $0 \leqslant y \leqslant \infty$ is unity, and the two functions are shown in Figure 2.10.3.

The propagation of a surface of constant ψ may be considered as follows. From equation (2.10.15) we see that this surface is located at

$$x = 2C\sqrt{(Dt)} \tag{2.10.17}$$

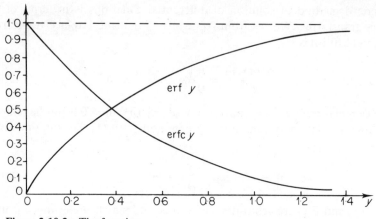

Figure 2.10.3 The function

$$\text{erf } y = \frac{2}{\sqrt{\pi}} \int_0^y e^{-\alpha^2} \, d\alpha$$

and the complementary function

$$\text{erfc } y = 1 - \frac{2}{\sqrt{\pi}} \int_0^y e^{-\alpha^2} \, d\alpha$$

where C is a constant, the field adopting the value erfc C. The velocity of propagation of this surface is, therefore,

$$\frac{\mathrm{d}x}{\mathrm{d}t} = C\sqrt{(D)}t^{-\frac{1}{2}} \tag{2.10.18}$$

and is seen to decrease with time.

As a further example of the use of separated equations in what at first sight seems an innocuous problem, consider the evolution of the temperature distribution $\psi(x, y, t)$ across the rectangular cartesian frame shown in Figure 2.10.4(a). The rectangle, initially at temperature zero, discontinuously has its boundary $x = 0$ maintained at a temperature Θ and the temperature distribution at $t = 0$ is described by $\psi(x, y, 0)$. These boundary conditions are maintained,

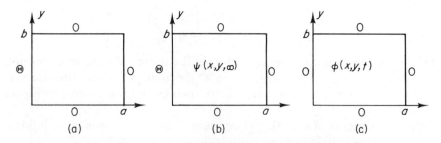

Figure 2.10.4 (From M. Ya. Azbel', M. I. Kaganov and I. M. Lifshitz, 'Conduction electrons in metals'. Copyright © 1973 by Scientific American, Inc. All rights reserved)

and the distribution eventually attains its steady state form $\psi(x, y, \infty)$, Figure 2.10.4(b). The steady-state $(\partial\psi/\partial t = 0)$ diffusion equation $\nabla^2\psi = 0$ is of Laplacian form and we obtain directly the separated equations

$$\frac{d^2 X}{dx^2} = m^2 X \tag{2.10.19}$$

$$\frac{d^2 Y}{dy^2} = -m^2 Y \tag{2.10.20}$$

where we assume $\psi(x, y) \equiv X(x)Y(y)$. Equation (2.10.20) has solutions satisfying the y boundary condition of the form:

$$Y(y) = \sin\left(\frac{n\pi y}{b}\right) \tag{2.10.21}$$

where $n = 0, 1, 2, \ldots$.

Equation (2.10.19) therefore has solutions satisfying the x boundary condition of the form:

$$X(x) = \sinh\left(\frac{n\pi(a - x)}{b}\right) \tag{2.10.22}$$

and so the steady state distribution is

$$\psi(x, y, \infty) = \sum_{n=0}^{\infty} a_n \sin\left(\frac{n\pi y}{b}\right) \sinh\left(\frac{n\pi(a - x)}{b}\right)$$

At $x = 0$, $\psi(x, y, \infty) = \Theta$. Multiplying throughout by $\sin(n\pi y/b)$ and integrating we obtain from the orthogonality property of the sine function:

$$a_n = 4\Theta\left[n\pi \sinh\left(\frac{n\pi a}{b}\right)\right]^{-1}$$

where n is restricted to odd integers if a_n is to be non-vanishing. Alternatively, we may allow n to adopt any integral value and write

$$\psi(x, y, \infty) = \frac{4\Theta}{\pi} \sum_{n=0}^{\infty} \frac{1}{(2n + 1)} \sin\left[\frac{(2n + 1)\pi y}{b}\right] \frac{\sinh\left[(2n + 1)\pi(a - x)/b\right]}{\sinh\left[(2n + 1)\pi a/b\right]}$$

$$\tag{2.10.23}$$

Now let us consider the problem represented by Figure 2.10.4(c). The boundaries are all permanently maintained at zero, and we enquire how the initial distribution $\phi(x, y, 0)$ evolves subject to the usual diffusion equation

$$\nabla^2\phi(x, y, t) = \frac{1}{D}\frac{\partial\phi(x, y, t)}{\partial t} \tag{2.10.24}$$

We may immediately write down the solution

$$\phi(x, y, t) = \sum_{p=0}^{\infty} \sum_{q=0}^{\infty} c_{pq} \sin\left(\frac{p\pi x}{a}\right) \sin\left(\frac{q\pi y}{b}\right) \exp\left\{-D\left[\left(\frac{p\pi}{a}\right)^2 + \left(\frac{q\pi}{b}\right)^2\right]t\right\}$$

$$\tag{2.10.25}$$

where $p = 0, 1, 2, \ldots, q = 0, 1, 2, \ldots$.

The coefficients c_{pq} may be determined as usual from the orthogonality condition at $t = 0$

$$c_{pq} = \frac{4}{pq} \int_0^a \int_0^b \phi(x, y, 0) \sin\left(\frac{p\pi x}{a}\right) \sin\left(\frac{q\pi y}{b}\right) dy\, dx \qquad (2.10.26)$$

We should notice, however, that the boundary conditions described by $\phi(x, y, t)$ (Figure 2.10.4(c)) are *also* satisfied by $\psi(x, y, 0) - \psi(x, y, \infty)$, i.e.

$$\phi(x, y, 0) = \psi(x, y, 0) - \psi(x, y, \infty) \qquad (2.10.27)$$

If we now substitute the expression (2.10.27) in (2.10.26) we may determine the coefficients c_{pq}. The final solution may therefore be written:

$$\psi(x, y, t) = \psi(x, y, \infty) + \phi(x, y, t) \qquad (2.10.28)$$

where $\psi(x, y, t)$ satisfies the boundary conditions throughout. Explicitly

$$\psi(x, y, t) = \frac{4\Theta}{\pi} \sum_{n=0}^{\infty} \frac{1}{(2n+1)} \sin\left[\frac{(2n+1)\pi y}{b}\right] \frac{\sinh\left[(2n+1)\pi(a-x)/b\right]}{\sinh\left[(2n+1)\pi a/b\right]}$$

$$+ \sum_{p=0}^{\infty} \sum_{q=0}^{\infty} c_{pq} \sin\left(\frac{p\pi x}{a}\right) \sin\left(\frac{q\pi y}{b}\right) \exp\left\{-D\left[\left(\frac{p\pi}{a}\right)^2 + \left(\frac{q\pi}{b}\right)^2\right]\right\}t$$

$$(2.10.29)$$

where c_{pq} is given by equation (2.10.26). We may regard the steady state solution as those terms in the expression (2.10.25) for which $(p/a)^2 + (q/b)^2 = 0$, thereby removing the time-dependence, and giving $p = \pm iq$ so that the harmonic solutions $\rightarrow$ hyperbolic solutions.

Problems

1 Sketch the general form of a typical intermolecular potential, and outline the reasons for believing that such a potential exists. A simple model of an ionic crystal is a line of alternately positively and negatively charged ions, each carrying a single electronic charge. The potential between any two ions separated by a distance r includes, in addition to the Coulomb interaction, a repulsive component of the form λ/r^{10}, where λ is a constant. The spacing between the ions in the chain is 4 Å. Calculate the energy of formation in electron-volts per ion of the crystal. The chain is immersed in water of dielectric constant 80. Assuming that the effect of this is to alter the electrostatic part of the binding energy, but to leave the repulsive part unchanged, calculate the new inter-ionic spacing and binding energy. Hence show that you expect the crystal to be soluble in water at room temperature.

$$[1 - \tfrac{1}{2} + \tfrac{1}{3} - \tfrac{1}{4} + \cdots] = 0{\cdot}69; \quad \sum_{n=1}^{\infty} \frac{1}{n^p} \text{ converges for } p < 1$$

2 Show that a wave travelling along the x-direction may be represented by an equation of the form

$$\frac{d^2y}{dx^2} = \frac{1}{V^2}\frac{d^2y}{dt^2}$$

Discuss, without detailed analysis, the behaviour of waves which meet a boundary between two media. Illustrate your discussion with particular reference to (i) compression waves in a fluid, (ii) electromagnetic waves, and (iii) waves associated with particles. A thin-walled metal tube has a length of 100 cm, a radius of 1 cm and a mass of 200 g, and is rigidly clamped at one end. When a couple is applied to the free end the torsion constant is found to be 10^8 dyn cm rad^{-1}. Find the lowest frequency of angular oscillation of the free end.

3 The plates of a parallel-plate condenser of capacitance C are charged to a potential difference V and held apart by suitable forces. Discuss the energy changes that occur when the separation of the plates is slowly increased by a small amount (a) when the plates are insulated, (b) when they are connected to a battery of e.m.f. V. Hence, or otherwise, find the force acting on a long cylinder of radius 1 cm, which hangs in a coaxial cylinder of the same length and radius 2 cm, if the end of one cylinder is near the centre of the other, and if a potential difference of 15,000 V is maintained between the cylinders.

What would be the force if the whole system were immersed in oil of dielectric constant 3?

4 Explain the method of images used in electrostatics. Show how it could in principle be applied to find the potential of a point charge q situated at the centre of a cubical box, made of conducting material and earthed.

Two point charges are placed on a common normal to a fixed infinite conducting plane and on the same side of it; one charge (Q) is twice as far from the plane as the other (q). Show that Q experiences no resultant force, provided that $Q = (128/9)q$. Would you expect the equilibrium of Q to be stable in these circumstances?

5 A thin elastic membrane of mass ρ per unit area is stretched into a plane surface by a uniform tension T, and it is then clamped along two straight parallel lines a distance d apart. Discuss the propagation of transverse waves along the strip of membrane that lies between the clamps, and show that no propagation is possible below a certain cut-off frequency.

In a particular case ρ is 1 gm cm^{-2}, T is 100 dyne cm^{-1} and d is 5 cm. If the membrane is oscillated at one end at a frequency of 2 cycles per second, how long will it take the disturbance to travel along the membrane to a point 100 cm away?

6 Explain carefully the operation of the coatings used on high quality lenses to reduce unwanted reflections.

If an electron with kinetic energy E is aimed at normal incidence at a sharp potential step, across which its potential energy rises abruptly from zero to a uniform value $V(< E)$, there is always a chance that it will be reflected at the step. Show, by analogy with the equivalent optical problem or otherwise, how the reflection can be eliminated by arranging that the potential rises to V in two steps rather than one, and calculate the height and width of the intermediate step required when E is 4 eV and V is 3 eV.

7 What is a laser? Explain briefly the way in which lasers function. A ruby crystal in the form of a cube with sides of 1 cm is well silvered on all its surfaces. Estimate the number of modes which might in principle be excited if the crystal were operated as a laser. Why is the number of modes excited in practice in a ruby laser relatively small?

(The ruby emission line has a wavelength, measured in vacuo, of 6943 Å and a width of 0·04 Å; refractive index of ruby = 1·76.)

8 Write down Schrödinger's equation for a particle in one dimension which experiences a potential V. Describe, with the aid of diagrams where appropriate, the nature of the solutions to this equation in the case where V is positive and proportional to x^2 (where x is linear displacement) but independent of time. Explain carefully the physical interpretation to be placed upon them. In particular, show that one solution exists in which the wavefunction is proportional to e^{-ax^2}, where a is a positive constant. Determine the energy corresponding to this solution and comment on your result.

9 Why is the differential equation

$$\ddot{x} + 2k\dot{x} + n^2 x = 0$$

of importance in physics? Discuss the different types of solution for it that may occur.

A sphere of radius 1 mm and uniform density $1 \cdot 5 \, \text{gm cm}^{-3}$ is suspended from a fixed support by a light spring. If the sphere is immersed in a fluid of density $1 \cdot 0 \, \text{gm cm}^{-3}$ its equilibrium position is found to rise by 1 cm. Calculate the period with which the sphere oscillates if it is slightly displaced from equilibrium in a vertical direction (a) in air, (b) when it is immersed in the fluid. The viscosity of the fluid is $0 \cdot 1 \, \text{gm cm}^{-1} \, \text{sec}^{-1}$. ($g = 1000 \, \text{cm}$ sec^{-2}. You may assume Stokes' formula, which says that the force opposing the motion of a sphere of radius a, travelling in the x-direction through a fluid of density ρ and viscosity η, is given for small velocities by $6\pi\eta a\dot{x} + \frac{2}{3}\pi a^3 \rho \ddot{x}$).

10 Heat is liberated at a rate $R(x, y, z, t)$ per unit volume inside a body by applying a very high frequency electric field. The material of the body has conductivity K, specific heat C per unit mass, and density ρ, and there is some flow of heat to the surface of the body where it is lost by cooling. Derive the partial differential equation satisfied by the temperature T inside the body, considered as a function of the time and position.

A large slab 3 cm thick, with $K = 0 \cdot 0012 \, \text{cal sec}^{-1} \, \text{cm}^{-1} \, \text{deg}^{-1}$, has heat liberated inside at a constant rate $R = 0 \cdot 104 \, \text{cal cm}^{-3} \, \text{sec}^{-1}$. If the outside is kept at a constant temperature of 20 °C, calculate the maximum temperature inside the slab when a steady state has been reached.

11 Indicate briefly the importance of the concept of standing waves in connection with *either* black body radiation *or* the theory of specific heat.

Determine the number of possible modes of vibration in a cube of aluminium of side 10 cm in a frequency range from 1×10^{11} c/s to $1 \cdot 0001 \times 10^{11}$ c/s. The velocities of longitudinal and transverse waves in aluminium are $6 \times 10^5 \, \text{cm sec}^{-1}$ and $3 \times 10^5 \, \text{cm}$ sec^{-1} respectively.

12 A particle moves in one dimension under the influence of a (smooth) potential $V(x)$. Show that if it is displaced slightly from a position of stable equilibrium it executes approximately simple harmonic motion. Explain how this generalizes to motion in more than one dimension and what is meant by normal modes. A uniform rod of length l and mass m has a bob, also of mass m, attached to one end by a light string of length l. The rod is pivoted at the other end so that the system may make small oscillations in a vertical plane. Show that the ratio of the frequencies of the two truly periodic motions is approximately 3·74 to 1. The rod is initially displaced by 10° from the vertical and when the system is released from rest it subsequently executes the faster of the two periodic motions. Find the initial inclination of the string and sketch the initial configuration.

13 A particle of mass m which moves in a one-dimensional potential well $V(x) = \frac{1}{2}kx^2$ is in the first excited state having the (normalized) wavefunction

$$\psi(x) = (2\alpha\pi^{-\frac{1}{2}})^{\frac{1}{2}}\alpha x \, e^{-\frac{1}{2}\alpha^2 x^2}$$

where $\alpha = (mkh^{-2})^{\frac{1}{4}}$.

(a) Prove that the mean (i.e. expectation) value of its potential energy is equal to the mean value of its kinetic energy.

(b) Hence show that if p is the momentum of the particle,

$$\langle x^2 \rangle \langle p^2 \rangle = \tfrac{9}{4}\hbar^2$$

where $\langle \ \rangle$ denotes mean value; comment on the physical significance of this result.

(c) Show that the probability that the particle will be found in a classically inaccessible region is equal to

$$4\pi^{-\frac{1}{2}} I_1(\sqrt{3})$$

where

$$I_n(0) = \int_r^\infty y^{2n} e^{-y^2} \, dy$$

$$\left(\text{You may need the formula } I_n(0) = \frac{1 \cdot 3 \cdot 5 \cdots (2n-1)\pi^{\frac{1}{2}}}{2^{n+1}} \right)$$

14 Write down the Schrödinger equation for a system of two non-interacting indistinguishable particles confined to a region by a potential and discuss the symmetry properties of the acceptable solutions.

Two spin $\tfrac{1}{2}\hbar$ indistinguishable Fermi particles, each of mass m, are confined by the totally reflecting walls of a cube of side a. What are the wavefunctions and total spin quantum numbers of the two lowest energy states?

If the number of such particles in the box is increased to ten, calculate the ground state energy of the system. (For this part of the question, the wavefunctions are *not* required.)

15 Prove that, if ψ is a solution of the time-dependent Schrödinger equation,

$$\frac{\partial}{\partial t}|\psi|^2 - \frac{i\hbar}{2m} \, \text{div}\, (\psi^*\nabla\psi - \psi\nabla\psi^*) = 0$$

and interpret this equation in the light of the law of conservation of particles.

A particle of mass m travelling in the positive x-direction with energy $E > U_0$ is incident on a potential barrier $U(x)$ of arbitrary shape, but for which

$$U(x) \to U_0 \quad \text{as } x \to \infty$$

and

$$U(x) \to 0 \quad \text{as } x \to -\infty$$

By applying the relation proved above to the solutions of the Schrödinger equation, show explicitly that the transmission coefficient T and the reflection coefficient R are connected by the relation

$$T + R = 1$$

16 What is meant by a 'normal mode'?

A horizontal string of negligible mass and length $3l$ is maintained under tension T. Equal masses m are attached at points l and $2l$ and are free to move in a horizontal plane. Find expressions for the steady motions of the masses if one end of the string is made to move sinusoidally in a transverse direction in this plane with a small amplitude a and angular frequency ω while the other end of the string remains still. Sketch graphs to illustrate the effect of varying ω.

17 The spectrum of radiation of a gas consisting of diatomic molecules shows a series of lines equally spaced in terms of frequencies. Show that these lines can be interpreted quantum mechanically as due to changes in the rotational motion of the molecule. The wavefunction for a molecule consisting of two atoms of equal mass has the form $\psi = \text{const. } \psi_l(\theta, \phi)\,e^{-r/a}$, where r is the inter-atomic distance and $\psi_l(\theta, \phi)$ is a (normalized) angular momentum eigenfunction associated with the quantum number l. Determine the rotational energy which corresponds to this wavefunction. The wavelengths of one pair of adjacent lines are $2\cdot15 \times 10^{-3}$ and $2\cdot85 \times 10^{-3}$ cm. Each atom of the gas has mass $3\cdot34 \times 10^{-24}$ g. Find the value of the parameter a, and correlate the two spectral lines with the corresponding l-values.

$$\left(\int_0^\infty r^n\,e^{-br}\,dr = \frac{n!}{b^{n+1}}\right.$$

where n is a positive integer and $b > 0$)

18 In an attempt to describe the motion of the protons and neutrons in an atomic nucleus it is assumed that each nucleon moves in an average potential of the form $V(r) = -V_0 + \frac{1}{2}kr^2$. Show that a state of a nucleon can be written as a product $u(x)v(y)w(z)$ of one-dimensional oscillator wavefunctions. Determine the degeneracy of the first three energy levels and their energy eigenvalues.

A neutron and a proton in a nucleus are each in the lowest energy level. Considering their motion in three dimensions, determine their mean square distance apart in terms of k and the mass m of a neutron or proton.

(Hint: the normalized ground state eigenfunction of the one-dimensional oscillator of mass m is

$$u_0 = \left(\frac{k}{\pi\hbar\omega}\right)^{\frac{1}{4}} \exp\left(-\frac{1}{2}\frac{kx^2}{\hbar\omega}\right)$$

where $\omega^2 = k/m$

$$\int_0^\infty r^{2n}\,e^{-ar^2}\,dr = \frac{1\cdot3\cdot5\cdots(2n-1)}{2^{n+1}a^n}\left(\frac{\pi}{a}\right)^{\frac{1}{4}}$$

19 Derive the differential equation

$$\nabla^2\theta - \frac{1}{D}\frac{\partial\theta}{\partial t} = 0$$

for the temperature θ inside a body. What is D? What is the correct form for the equation when the thermal conductivity K, the specific heat C, and density ρ are allowed to vary with position?

A long thermally insulated rod of length L and cross-sectional area A has constant thermal conductivity K. At one end is wound a small resistance coil through which is passed an electric current $I_0 \cos \Omega t$, the heat generated going into the rod. The other end is maintained cold. Calculate the velocity of the temperature wave which is propagated down the rod, and its attenuation per unit length.

20 Derive expressions for the quantum mechanical wavefunctions and the energy levels for a particle confined in a rectangular box with sides L_x, L_y, L_z. The potential inside the box may be assumed to be zero and the walls to be perfectly reflecting.

A cubical box of this nature may be divided into two equal parts with a diaphragm parallel to two of the faces. Calculate the ratio of the lowest possible energies of the system when (a) the diaphragm is in position with three electrons on each side of it, and (b) when the diaphragm is removed, and the six electrons can move freely in the whole box.

21 In a sodium chloride crystal, the atomic environment round a sodium ion has full cubic symmetry. It is desired to represent by an algebraic expression the electric potential produced in the region of a sodium ion by the neighbouring ions.

Giving the relevant theory, test the following expressions to find which have the correct symmetry and satisfy the equations necessary for an electrostatic potential Ar^2; $B[x^4 + y^4 + z^4 - (\frac{3}{5}r^4)]$; $Cxyz$; where A, B and C are constants.

22 A collection of particles is defined by a wavefunction ψ. From the fact that the number of particles is constant, show that the flux of particles may be taken to be

$$\mathbf{j} = (\hbar/2im)(\psi\nabla\psi^* - \psi^*\nabla\psi)$$

A beam of non-interacting particles of electronic mass moves in space. The spatial part of the wavefunction is

$$\psi = A\,e^{ik_1x}\,e^{ik_2y}$$

where A, k_1 and k_2 are constants. The energy of each particle in the beam is 1000 eV and the observed particle fluxes along the x- and y-axes are 5×10^8 and 3×10^8 cm^{-2} sec^{-1} respectively. Calculate the two wave numbers k_1 and k_2 and the probability of finding a particle per unit volume.

23 An atom of mass m oscillates in one dimension in a region where the potential energy $V = kx^2$, k being constant. Write down the time-independent Schrödinger equation for the motion and verify that one possible solution of the equation is

$$\psi = A[2(x/a)^2 - 1]\exp[-x^2/2a^2]$$

where A and a are constants.

Calculate the value of a and the energy of the state in terms of k and m.

Determine also at what distance from the centre the probability of finding the particle (per unit distance) is a maximum. Is this equal to the classical amplitude of oscillation? If the former distance is 10^{-8} cm and the atom is a hydrogen atom, calculate the energy of the state in ergs.

CHAPTER 3

Orthogonal Curvilinear Coordinate Systems: Cylindrical Solutions

3.1 Introduction

As we observed in the previous chapter, a classification of the solutions to the various field equations is conveniently made in terms of the symmetry of the coordinate system. Consequently we found that in the rectangular cartesian frame the solutions were generally products of simple trigonometric functions. Moreover, perturbation of the normal mode solutions could still be described in terms of linear combinations of normal mode solutions. This, as we shall shortly see, is by no means specific to the cartesian frame, and the expression of the field in terms of complete orthonormal sets of eigenfunctions of the appropriate symmetry will become as familiar as the Fourier expansions in rectangular coordinates.

In this chapter we shall consider solutions of the field equations in cylindrical polar coordinates in which the field is completely specified by the distribution $\psi(r, \phi, z)$ (Figure 3.1.1).

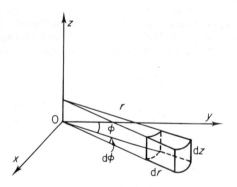

Figure 3.1.1 The volume element in cylindrical coordinates

The Laplacian of the potential field $\psi(r, \phi, z)$ in cylindrical coordinates is

$$\nabla^2 \psi(r, \phi, z) = \frac{\partial^2 \psi}{\partial r^2} + \frac{1}{r} \frac{\partial \psi}{\partial r} + \frac{1}{r^2} \frac{\partial^2 \psi}{\partial \phi^2} + \frac{\partial^2 \psi}{\partial z^2}$$

and in the homogeneous case of Laplace's equation $\nabla^2 \psi = 0$, we may try the separable product

$$\psi(r, \phi, z) = R(r)\Phi(\phi)Z(z)$$

and substitute in Laplace's equation to give

$$\Phi Z \frac{d^2R}{dr^2} + \frac{\Phi Z}{r} \frac{dR}{dr} + \frac{RZ}{r^2} \frac{d^2\Phi}{d\phi^2} + R\Phi \frac{d^2Z}{dz^2} = 0 \qquad (3.1.1)$$

Division throughout by $R\Phi Z$ enables us to separate the Z-equation:

$$\frac{d^2Z}{dz^2} - k^2 Z = 0 \qquad (3.1.2)$$

where k^2 is the separation constant, and equation (3.1.1) becomes upon multiplying throughout by r^2

$$\frac{r^2}{R} \frac{d^2R}{dr^2} + \frac{1}{R} \frac{dR}{dr} + \frac{1}{\Phi} \frac{d^2\Phi}{d\phi^2} + k^2 r^2 = 0 \qquad (3.1.3)$$

whereupon we may immediately separate the Φ-equation:

$$\frac{d^2\Phi}{d\phi^2} + m\Phi = 0 \qquad (3.1.4)$$

where m^2 is the separation constant, leaving the radial equation:

$$\frac{d^2R}{dr^2} + \frac{1}{r} \frac{dR}{dr} + \left(k^2 - \frac{m^2}{r^2} \right) R = 0 \qquad (3.1.5)$$

If there is no z-dependence, that is $Z(z) = $ constant, then $k = 0$ and the solutions are effectively two-dimensional, $\Phi(r, \phi)$. The Z-equation has trigonometric or hyperbolic solutions according as k is imaginary or real, and this will of course correspondingly affect the form of the radial equation. If k is real then we generally obtain two sorts of radial solution $J_m(kr)$ and $Y_m(kr)$ termed the Bessel functions of the first and second kinds, whilst if k is imaginary the hyperbolic Bessel functions K_m and I_m are obtained. The angular equation in ϕ results in trigonometric solutions of the form $\sin m\phi$, $\cos m\phi$, and the condition that the solution be continuous at $\phi = 0, 2\pi$, restricts the values of m to integers, $m = 0, 1, 2, 3, \ldots$.

The Bessel functions form an orthogonal set, and rival the trigonometric functions in their versatility and applicability to physical problems. We perhaps might anticipate that *spherical* Bessel functions of the various kinds may also develop, but we shall defer further discussion until the next chapter.

The development of Laplace's equation in cylindrical coordinates given above has been made without specification of the physical nature of the field, i.e. whether it is electrostatic, magnetic, thermal, probabilistic, etc. The only condition specified is that there should be no distribution of sources or sinks within the region other than that at the origin. The detailed form of the separated equations (3.1.5) will differ from problem to problem in the inhomogeneous (Poisson) case when there is a source–sink distribution in the region. Indeed, it may well be that no separation of the type outlined above is possible, and approximate methods of solution are enforced.

The problem of the infinitely long, uniform line source represents a particularly simple application of Laplace's equation in cylindrical coordinates. If we take the field ψ to be purely a function of the radial distance from the source, and not a function of polar angle ($d\Phi/d\phi = 0$) then the Laplace equation reduces to

$$\nabla^2 \psi(r) = \frac{d^2 R}{dr^2} + \frac{1}{r}\frac{dR}{dr}$$

$$= \frac{1}{r}\frac{d}{dr}\left(r\frac{dR}{dr}\right) = 0 \tag{3.1.6}$$

Solving for $\psi(r)$ yields

$$\psi(r) = A \log_e \frac{1}{r} + B \tag{3.1.7}$$

where A and B are arbitrary constants to be determined from the physical boundary conditions. Notice that equation (3.1.7) represents the solution in all regions except on the cylindrical axis, $r = 0$. As another example we may consider the electrostatic field between two concentric infinitely long cylinders

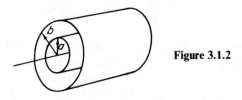

Figure 3.1.2

of radius a and b, uniformly charged to $\pm\lambda$ per unit length, Figure 3.1.2. If the potential of cylinder a is ψ_a and that of b, ψ_b, then

$$\psi_a = A \log_e \frac{1}{a} + B$$

$$\psi_b = A \log_e \frac{1}{b} + B \tag{3.1.8}$$

where $A = \lambda/2\pi\varepsilon$. The capacitance of the system, per unit length, $C = Q/V$ is

$$C = \lambda/(\psi_b - \psi_a) = 2\pi\varepsilon \log_e\left(\frac{b}{a}\right) \tag{3.1.9}$$

We could, of course, have arrived at the same conclusion using Gauss' theorem, and indeed the Laplace and Poisson equations represent differential forms of Gauss' theorem. We may also consider the field distribution in the source–sink configuration consisting of two uniform, infinite parallel line sources of density λ and $-\lambda$ (corresponding to a sink), separated by the distance

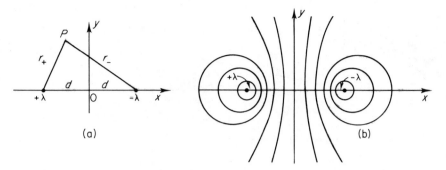

Figure 3.1.3 (a) The geometry of the parallel line charges, and (b) the associated equipotential distribution

$2d$, Figure 3.1.3(a). It follows from our previous discussion that the potential field ψ at the point P will be

$$\psi = \frac{\lambda}{2\pi\varepsilon}\ln\left(\frac{r_-}{r_+}\right)$$

$$= \frac{\lambda}{2\pi\varepsilon}\ln\sqrt{\frac{y^2+(x-d)^2}{y^2+(x+d)^2}}$$

(3.1.10)

in rectangular cartesian coordinates. The locus of an equipotential, i.e. for $\psi = $ constant, is given by

$$\psi_c = \frac{\lambda}{4\pi\varepsilon}\ln\frac{y^2+(x-d)^2}{y^2+(x+d)^2}$$

or,

$$y^2+(x-d)^2 = [y^2+(x+d)^2]\exp\left(\frac{4\pi\varepsilon\psi_c}{\lambda}\right)$$

(3.1.11)

which on rearrangement yields the cartesian equation for the family of circles

$$\left(x+d\coth\frac{2\pi\varepsilon\psi_c}{\lambda}\right)^2 + y^2 = d^2\left(\coth^2\frac{2\pi\varepsilon\psi_c}{\lambda}-1\right)$$

(3.1.12)

with centres at $[-d\coth(2\pi\varepsilon\psi_c/\lambda), 0]$ and radii $d[\coth^2(2\pi\varepsilon\psi_c/\lambda)-1]^{\frac{1}{2}}$. When the equipotential surface $\psi_c < 0$ in the vicinity of the sink the circles appear on the right of the neutral plane corresponding to $\psi_c = 0$ (Figure 3.1.3(b)). Similarly for $\psi_c > 0$, the circles appear on the left.

The interaction between a uniform line source of potential density λ and an infinite, parallel equipotential cylindrical surface of radius R may be also considered as follows.

We suppose that for the purposes of the interaction the cylindrical equipotential surface may be reproduced by an infinite line source of density $-\lambda$,

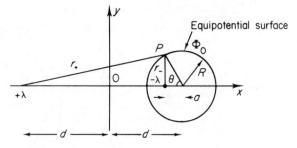

Figure 3.1.4 The geometry of the 'image' solution of
the interacting line charge and cylinder

located of course on the plane of centres (Figure 3.1.4). This merely amounts to choosing one of the family of equipotential surfaces discussed above to coincide with the cylindrical surface. Then the potential at a point P due to the line source $+\lambda$ and the hypothetical sink $-\lambda$ is

$$\frac{\lambda}{2\pi\varepsilon} \ln \frac{1}{r^+} - \frac{\lambda}{2\pi\varepsilon} \ln \frac{1}{r^-} = \psi_0 \tag{3.1.13}$$

i.e.

$$\left(\frac{r_-}{r_+}\right)^2 = \frac{R^2 + a^2 - 2aR\cos\theta}{R^2 + 4d^2 - 8dR\cos\theta}$$

$$= \exp\left(\frac{4\pi\varepsilon\psi_0}{\lambda}\right) \tag{3.1.14}$$

Cross-multiplication gives

$$R^2 + a^2 - 2aR\cos\theta = (R^2 + 4d^2 - 8dR\cos\theta)\exp\left(\frac{4\pi\varepsilon\psi_0}{\lambda}\right),$$

and comparison of terms in $\cos\theta$ gives

$$2aR = 8dR\exp\left(\frac{4\pi\varepsilon\psi_0}{\lambda}\right)$$

i.e.

$$a = 4d\exp\left(\frac{4\pi\varepsilon\psi_0}{\lambda}\right) \tag{3.1.15}$$

as the location of the 'image' sink line of density $-\lambda$. Thus the configuration in Figure 3.1.4 may indeed be replaced by the interaction of two parallel infinite line charges of density $\pm\lambda$. Further consideration of Figure 3.1.3(b) suggests that we may treat the interaction between two infinitely long parallel uniformly charged cylinders in the same way. Having obtained these satisfactory solutions, the theorem of uniqueness enables us to assert that these represent the only solution. This is, of course, an example of the method of images, and the analogous process for systems of spherical symmetry will be considered in the next chapter.

We may alternatively consider the solution of the Laplace or Poisson field equations when the potential is a function of the coordinate ϕ only, and in doing so a different set of relations for the potential and field intensity are obtained. The infinite line source in this case is known as a *vortex filament* or *vortex source*, and arises in the discussion of fluid fields and magnetic fields associated with an electrical current. The associated field is directed *around* the vortex source rather than radially away from it as in the previous case, and the vortex field **F** corresponds to the flux or fluid velocity in the direction $\hat{\phi}$ at the point r. For regions other than on the vortex filament itself, Laplace's equation holds and we have for the purely angle-dependent potential field

$$\nabla^2 \psi = \frac{d^2 \psi}{d\phi^2} = 0 \tag{3.1.16}$$

Solving for $\psi(\phi)$, we obtain

$$\psi(\phi) = -A\phi + B \tag{3.1.17}$$

If we further set the zero of potential such that $\psi(0) = 0$, then it follows that $B = 0$. The associated field is given as

$$
\begin{aligned}
\mathbf{F} &= -\nabla\psi \\
&= -\left(\hat{\mathbf{r}}\frac{\partial}{\partial r} + \frac{\hat{\phi}}{r}\frac{\partial}{\partial \phi}\right)\psi(\phi) \\
&= \frac{A}{r}\hat{\phi}
\end{aligned} \tag{3.1.18}
$$

in cylindrical coordinates, where $\hat{\mathbf{r}}$ and $\hat{\phi}$ are the unit radial and angular vectors, respectively. The flux density, or fluid velocity, associated with the field in a homogeneous and isotropic system is of course proportional to the vortex field, **F**. The circulation, Γ, associated with the vortex field is given by the integral

$$\Gamma = \oint_c \mathbf{F} \cdot d\mathbf{l} = \oint_A \text{curl } \mathbf{F} \, dA$$

where the second integral is taken over the area enclosed by the contour c in the first. If the contour does not enclose the vortex source, then

$$\text{curl } \mathbf{F} = 0$$

where curl **F** is termed the *vorticity* of the fluid, and **F** is identified as a vector potential. If, however, the vortex filament is enclosed by the contour, we have from Stokes' theorem

$$\Gamma = \int_0^{2\pi} \frac{A\hat{\phi}}{r}r \, d\hat{\phi} = \oint \text{curl } \mathbf{F} \, dA \tag{3.1.19}$$

where we have taken a circular contour in the (x, y) plane about the vortex source, assumed to lie on the z-azis. The vortex field may therefore be written

$$F = \frac{\Gamma \hat{\phi}}{2\pi r}$$

To fix our ideas we can determine the vortex potential associated with an infinitely long straight current-carrying conductor, the vortex source being directed along the wire as shown in Figure 3.1.5. If the wire carries a current i

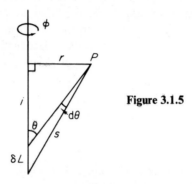

Figure 3.1.5

then the field at a point P due to current element δL is, according to the Biot–Savart law

$$\delta F = \frac{i}{4\pi} \frac{\delta L \times \hat{s}}{\hat{s}^3}$$

where $\hat{s}$ is the unit vector. The total field at P due to the whole wire is evidently

$$F = \frac{i}{4\pi r} \int_0^\pi \sin \theta \, d\theta \, \hat{\phi} = \frac{i}{2\pi} \frac{\hat{\phi}}{r} \tag{3.1.20}$$

where we have regained the original relation (3.1.18) between the vortex source and its associated field. We may now identify ψ as the *magnetic scalar potential*, and F as the *magnetic field* H associated with the current i.

The relation corresponding to equation (3.1.19) becomes, in conjunction with equation (3.1.20),

$$\frac{i}{2\pi} \int_0^{2\pi} \frac{\hat{\phi}}{r} r \, d\hat{\phi} = i = \oint_A J \, dA = \oint_A \text{curl } B \, dA \tag{3.1.21}$$

where J is the total current density enclosed by the contour. Since this result holds for any area and the contour enclosing it, we obtain

$$\text{curl } B = J \tag{3.1.22}$$

which is a statement of Ampère's law. We may alternatively identify J as the vorticity of the field.

An example of the development of a vortex filament arises in the discussion of hydrodynamic flow about an elliptical aerofoil (Section 3.8).

The generation of quantized vortices in superfluid liquid helium II has been proposed as a mechanism for the sudden development of a finite viscosity at a critical flow velocity. As an elementary excitation in the fluid it is clear that the vortex states are discontinuously excited beyond a critical velocity and below that the superfluid has its usual frictionless character, the ground state bosons being insufficiently excited to participate in the momentum transfer involved in viscous flow. A number of oscillating disc and sphere experiments have shown that beyond a critical amplitude of oscillation (corresponding of course to a critical angular velocity) the damping on the oscillator increases abruptly and this is ascribed to turbulence or the initiation of quantized vortex motions in the fluid.

Feynmann has made an order of magnitude calculation as follows.

We consider a stream of vortices to be emitted from a slit of width d as shown in Figure 3.1.6. The spacing of the vortices is x and the radius of the hole in the centre is a. The energy per unit length of one of these vortices is

$$E = \tfrac{1}{2} \int_a^d 2\pi r \rho v^2(r) \, \mathrm{d}r$$

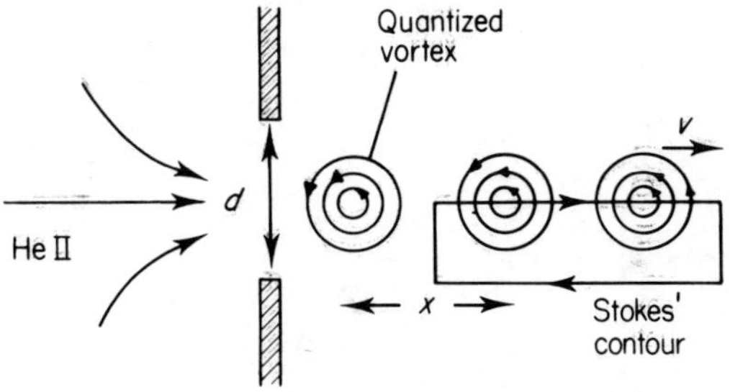

Figure 3.1.6

The classical velocity field $v(r)$ is given by equation (3.1.18), except that here we use the quantum mechanical representation:

$$E = \pi\rho \int_a^d r\left(\frac{h}{mr}\right)^2 \, \mathrm{d}r = \frac{\rho}{4\pi}\left(\frac{h}{m}\right)^2 \ln\left(\frac{d}{a}\right)$$

The number of vortex lines per unit length n may be determined by applying Stokes' theorem along a contour of unit length in the flow direction, returning outside the stream (Figure 3.1.6):

$$v1 = \frac{h}{m}n$$

But $n = 1/x$ and so

$$v = h/mx$$

v/x lines are created per second, and so the energy per second involved in vortex creation will be

$$\frac{v}{x}\pi\rho\left(\frac{\hbar}{m}\right)^2 \ln\left(\frac{d}{a}\right)$$

The energy per unit volume of the fluid flow is $\frac{1}{2}\rho v^2$, and so if we equate the flow energy per second to the rate of vortex energy production

$$\frac{v^2\pi\rho(\hbar/m)^2 \ln (d/a)}{\hbar/m} = \frac{1}{2}\rho v^2 vd$$

we have as a lower estimate of the critical velocity v_c required to excite quantized vortices:

$$v_c = \frac{\hbar}{md} \ln\left(\frac{d}{a}\right)$$

Taking a slit width of $d \sim 10^{-4}$ cm and a core radius $\sim 3 \times 10^{-8}$ cm, we find for the critical velocity

$$v_c \sim 100 \text{ cm/sec}$$

According to Feymann's expression the critical velocity diminishes as the slit width increases and this appears in qualitative agreement with Atkin's measurements shown in Figure 3.1.7.

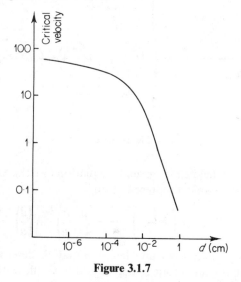

Figure 3.1.7

3.2   Wave Equation for a Uniform Stretched Circular Membrane

A particularly important problem which recurs in several branches of physics is that of the normal modes of vibration of a uniform stretched circular membrane, rigidly held around its circular perimeter. In this case the wave equation

(equation (1.4.3))

$$\nabla^2 \psi(r, \phi, t) = \frac{1}{c^2} \frac{\partial^2 \psi}{\partial t^2}(r, \phi, t)$$

becomes

$$\frac{\partial^2 \psi}{\partial r^2} + \frac{1}{r} \frac{\partial \psi}{\partial r} + \frac{1}{r^2} \frac{\partial^2 \psi}{\partial \phi^2} = \frac{1}{c^2} \frac{\partial^2 \psi}{\partial t^2} \tag{3.2.1}$$

Assuming that the solution may be written as the separable product

$$\psi(r, \phi, t) = R(r)\Phi(\phi)T(t) \tag{3.2.2}$$

then we obtain first of all

$$T(t) = A \sin(\omega t + \delta) \tag{3.2.3}$$

where ω is the frequency of vibration and δ is a phase constant. Equation (3.2.1) may now be written

$$\frac{\partial^2 \psi}{\partial r^2} + \frac{1}{r} \frac{\partial \psi}{\partial r} + \frac{1}{r^2} \frac{\partial^2 \psi}{\partial \phi^2} + \frac{\omega^2 \psi}{c^2} = 0 \tag{3.2.4}$$

and further separation yields the final pair of differential equations

$$\frac{d^2 R}{dr^2} + \frac{1}{r} \frac{dR}{dr} + \left[k - \left(\frac{n}{r} \right)^2 \right] R = 0 \tag{3.2.5}$$

$$\frac{d^2 \Phi}{d\phi^2} + n^2 \Phi = 0 \tag{3.2.6}$$

where n^2 is a constant and $k = \omega^2/c^2$. The condition that Φ be everywhere single valued requires that

$$\Phi(\phi) = \Phi(\phi + 2\pi) \tag{3.2.7}$$

and therefore that n be an integer: $0, \pm 1, \pm 2, \pm 3 \ldots$. Solutions to equation (3.2.6) are of the simple trigonometric form

$$\Phi(\phi) = C_n \sin n\phi + D_n \cos n\phi \tag{3.2.8}$$

whilst the radial solutions to equation (3.2.5) are termed the *Bessel functions of the first $J_n(x)$ and second $Y_n(x)$ kinds* where n is a positive or negative integer. J_{-n} is simply $(-1)^n J_n$, and we must look further for a second independent solution, this being Y_n. The general solution to equation (3.2.4) will therefore be of the form

$$\psi(r, \phi) = R(r)\Phi(\phi)$$

$$= \sum_{m=0}^{\infty} \sum_{n=0}^{\infty} [A_{mn} J_n(k_m r) + B_{mn} Y_n(k_m r)](C_n \sin n\phi + D_n \cos n\phi) \tag{3.2.9}$$

where k_m and n refer to the separation constants arising in equations (3.2.5), (3.2.6). The arbitrary coefficients A_{mn}, B_{mn}, C_n and D_n are to be determined as usual from the boundary conditions. A few low-order Bessel functions of the first and second kinds are shown in Figure 3.2.1. Those of the second kind are seen to be quite distinct from their positive-integer counterparts, and indeed they are characterized by being unbounded at the origin. In the solution of

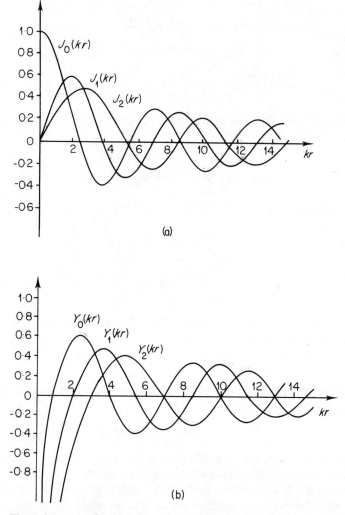

Figure 3.2.1 (a) The Bessel functions of the first kind, $J_n(kr)$. Notice that only $J_0(kr)$ is non-zero at the origin, although all are nevertheless finite. (b) The Bessel functions of the second kind $Y_n(kr)$. Notice that all these functions are negatively infinite at the origin, and in consequence are generally rejected in the solution of interior physical problems

interior physical problems the boundary requirements almost invariably require us to set $B_{mn} = 0$ in equation (3.2.9).

The constant k is seen to enter as a scaling parameter: the greater k the more compressed are the oscillations into the origin. The full solution of Bessel's equation (3.2.5) is given in Appendix III together with some of the important properties of the Bessel functions.

To return to the discussion of the normal modes of a circular membrane of radius a, we see that since each of the functions repeatedly passes through zero as the argument increases then we can satisfy the boundary condition $\psi(a, \phi) = 0$ provided

$$J_n(k_m a) = 0 \tag{3.2.10}$$

Thus for a membrane of given radius only certain values of k are permitted: this is directly analogous to the fitting of the trigonometric functions to the boundary conditions in the case of the square membrane. Equation (3.2.10) physically requires that a node of the Bessel function should be located at $r = a$. The normal mode frequencies of the membrane are then related to the discrete values of k_m satisfying equation (3.2.10) by

$$v_{nm} = \frac{c}{2\pi}\sqrt{k_m} \tag{3.2.11}$$

from an earlier relation (equations (3.2.4), (3.2.5) and (3.2.6)). Here v_{nm} is taken to represent the vibrational frequency of the 'mth harmonic' of the nth-order Bessel function of the first kind. The radial profile of the first few vibrational modes is shown in Figure 3.2.1(a) where the centre of the membrane coincides with the origin and the perimeter with the mth crossing of the radial axis. It is quite apparent that there is no simple harmonic relationship between the fundamental and overtones of the membrane. For this reason instruments based on two-dimensional vibrating systems produce little more than noise. However, we may observe that at large r the second and fourth terms in Bessel's equation (3.2.5) may become sufficiently small that the equation reduces to trigonometric form. Indeed, at large r

$$J_n(kr) = \sqrt{\frac{2}{\pi k r}} \sin\left(kr - (n + \tfrac{1}{2})\frac{\pi}{2}\right)$$

and by damping out the central vibrations with his hand a musician may approximately 'tune' his instrument.

This asymptotic form of the Bessel function has been encountered before in our discussion of highly excited normal modes of vibration of a freely hanging chain (Section 2.4).

If we now include the angular solutions it would seem that for given m and n there may be any number of possible modes of vibration according to the values of C_n and D_n. However only two of these modes are linearly independent

$$\psi(r, \phi) = CJ_n\left(\frac{2\pi v_{nm}}{c}\right) \cos n\phi$$

$$\psi(r, \phi) = CJ_n\left(\frac{2\pi v_{nm}}{c}\right) \sin n\phi$$

(3.2.12)

and it therefore follows that all solutions, except for $n = 0$, are doubly degenerate.

The normal modes of vibration may be particularly elegantly investigated by holographic techniques. The experimental layout is indicated schematically in Figure 3.2.2. A coherent monochromatic beam of light emitted from the laser is

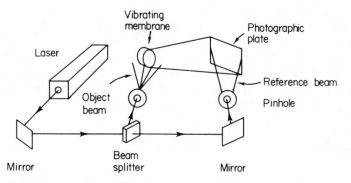

Figure 3.2.2 The experimental arrangement for the production of a hologram. The narrow laser beam is made divergent by the pinholes thereby illuminating the vibrating membrane and photographic plate. Since both the object and the reference beam are precisely phase related, phase differences in the two beams at the plate contain information on the profile of the membrane

split by a semi-silvered mirror or 'beam-splitter' into an object and a reference beam. These beams on passing through a pinhole aperture diverge into two beams precisely phase-related by the difference in the optical path each has travelled. The 'reference' beam passes straight onto the photographic plate whilst the object beam scatters off the vibrating membrane which is assumed to be executing a normal stationary mode. Clearly the phase surface of the scattered wavefront will be modified by the profile of the scattering surface, and interference processes will occur in the region where the scattered and reference beams overlap. The photographic plate is therefore covered with very fine highly complex interference fringes, this being termed a *hologram*.

If the plate is subsequently viewed in the laser beam as if it were a diffraction grating, then the *reconstruction process* occurs and three-dimensional images exhibiting parallax are obtained. Of particular interest to us are the various normal mode patterns which the membrane may be shown to execute, and some of these are shown in Figure 3.2.3.

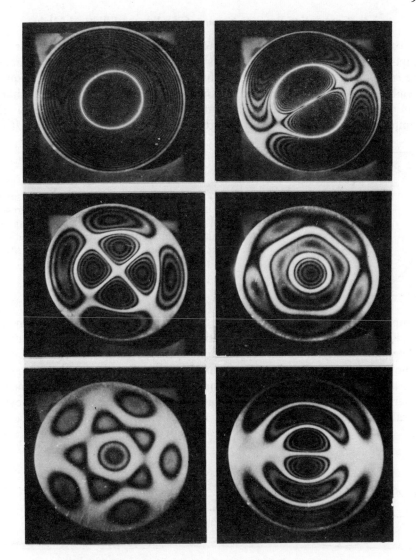

Figure 3.2.3 Modes of vibration of a circular membrane.

Top: (left) $J_0(k_1r)$, (right) $J_1(k_1r)\cos\phi$
Middle: (left) $J_2(k_1r)\cos 2\phi$, (right) $AJ_5(k_0r)\cos 5\phi + BJ_0(k_2r)$
Bottom: (left) $CJ_5(k_0r)\cos 5\phi + DJ_0(k_2r)$, (right) $J_1(k_2r)\cos\phi$.

Notice how the radial and angular modes associate: $J_m(kr)\cos m\phi$. The frequencies of the normal modes $J_5(k_0r)$ and $J_0(k_2r)$ are almost identical, and therefore hybridize extensively. (K.S. Pennington and K. A. Stetson, *Scientific American*, February 1968. Reproduced by permission of the authors)

3.3 Correlation Between Square, Circular and Elliptical Modes

A comparison between the normal mode vibrations for square, circular and elliptical membranes is particularly instructive since it enables us to establish the *correlation diagram* between the three systems. We may imagine a square membrane executing a normal vibrational mode, and investigate the modification as the system is slowly forced into one of the cylindrical modes as the damping in the shaded areas of Figure 3.3.1 is progressively increased. If the

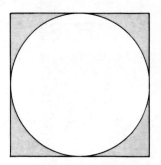

Figure 3.3.1 The adiabatic transformation of a square membrane into one of circular symmetry by progressive damping of the shaded areas. If the damping is applied slowly so that the appropriate normal mode is able to develop then a correlation between the circular and square normal modes may be established

damping is applied slowly, that is to say *adiabatically*, then the system will remain in a normal mode, and a correlation between the cartesian and cylindrical modes may be established. Application of the perturbation will in general remove the degeneracy of the cartesian modes, and in a few simple cases we may anticipate the correlation. For example, we should expect the fundamental square mode ($n = 1, m = 1$) to correlate directly with the fundamental cylindrical mode ($n = 0, m = 1$), Figure 3.3.2. Similarly we might expect the

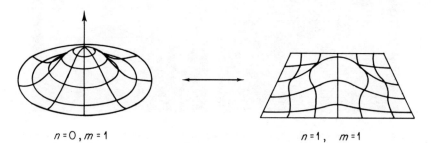

$$n = 0, m = 1 \qquad\qquad n = 1, \quad m = 1$$

Figure 3.3.2 The fundamental normal modes of a square ($n = 1, m = 1$) and a circular ($n = 0, m = 1$) membrane

degenerate square modes $(1, 2)$ $(2, 1)$ to correlate with the degenerate $(1, 1)$ cylindrical modes, Figure 3.3.3, although we note in this case that the degeneracy is preserved. Kauzmann has calculated the normal modes of the square and circular membranes, and has established the correlation diagram between the two systems, Figure 3.3.4. The relationships between the majority of the normal modes in the cylindrical and cartesian systems are, however, somewhat obscure.

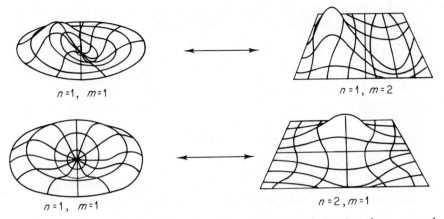

Figure 3.3.3 The degenerate first harmonic normal modes of vibration of a square and a circular membrane

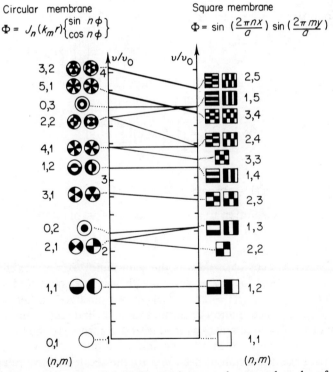

Circular membrane

$$\Phi = J_n(k_m r)\begin{Bmatrix}\sin n\phi \\ \cos n\phi\end{Bmatrix}$$

Square membrane

$$\Phi = \sin\left(\frac{2\pi nx}{a}\right)\sin\left(\frac{2\pi my}{a}\right)$$

Figure 3.3.4 Correlation diagram between the normal modes of vibration of a square and circular membrane. Note the extensive degeneracy which still persists, although the degeneracies are in some cases eliminated in one system with respect to the other. (W. Kauzmann, *Quantum Chemistry*, 1957. Reproduced by permission of Academic Press, Inc.)

Some rationalization may be effected by considering hybrid cartesian modes, such as those shown in Figure 2.6.2. In these cases the degeneracy is invariably destroyed.

The further perturbation of the cylindrical solutions by the imposition of an *elliptical* boundary condition is interesting. In cylindrical coordinates we may only consider boundaries of small ellipticity; a full solution may be developed in elliptical coordinates, but we shall not consider these solutions here (see Section 3.10).

An approximation to an elliptic boundary is given by

$$\xi = r(1 + \alpha \cos^2 \phi) \tag{3.3.1}$$

where the constant α is a measure of the ellipticity, and is assumed to be small, then the radial equation may be written, to a first approximation,

$$(1 + \alpha \cos^2 \phi)^2 \left\{ \frac{d^2 R}{d\xi^2} + \frac{1}{\xi} \frac{dR}{d\xi} \right\} + \left\{ k^2 - \frac{n^2}{\xi^2} (1 + \alpha \cos^2 \phi)^2 \right\} R = 0 \tag{3.3.2}$$

If terms of order α^2 are neglected, then equation (3.3.2) may be rewritten

$$\frac{d^2 R}{d\xi^2} + \frac{1}{\xi} \frac{dR}{d\xi} + \left\{ k^2 (1 - 2\alpha \cos^2 \phi) - \left(\frac{n}{\xi} \right)^2 \right\} R = 0 \tag{3.3.3}$$

Comparison of this equation with equation (3.2.5) shows that the final field distribution will consequently be of the form

$$\psi(\xi, \phi) = A_n J_n(\kappa_m \xi) \begin{pmatrix} \sin n\phi \\ \cos n\phi \end{pmatrix} \tag{3.3.4}$$

where

$$\kappa_m^2 = k_m^2 (1 - 2\alpha \cos^2 \phi) = 4\pi^2 v_{nm}^2 (1 - 2\alpha \cos^2 \phi)/c^2$$

We now require the solution to satisfy the boundary condition

$$J_n(\kappa_m \xi_a) = 0$$

where $\xi_a = a(1 + \alpha \cos^2 \phi)$, and a is the semi-minor axis of the ellipse. If we consider the elliptical locus $\xi = $ constant, then it is evident that κ will vary around that locus as $k\sqrt{(1 - 2\alpha \cos^2 \phi)}$. As we observed earlier, κ acts as a radial scaling parameter, and the normal mode cylindrical solutions are seen to be radially compressed and expanded with ϕ according to the above relation. As α tends to zero so we regain the cylindrical solutions, as we would expect. To anticipate the discussion in Section 3.10, the result we have obtained is in fact very close to the rigorous solution in elliptic coordinates. The extensive double-degeneracy of the normal mode cylindrical solutions is clearly removed in the case of an elliptical membrane in precisely the same way that degeneracy in the square membrane is removed in the rectangular case. A few of the simpler degenerate modes for the square and circular membrane together with their

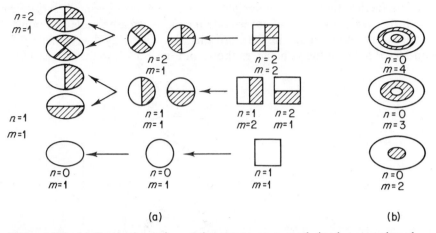

Figure 3.3.5 (a) Comparison of some degenerate square and circular normal modes. Application of an elliptical boundary condition eliminates degeneracy in many cases. (b) The first few purely radial modes for an elliptical membrane. The radial functions are related to the Bessel functions and are discussed in Section 3.10

non-degenerate elliptical counterparts are shown in Figure 3.3.5(a). The first few purely radial ($m = 0$) distributions (Figure 3.3.5(b)) are seen to exhibit an alternate compression and expansion of the Bessel function along the radius vector as it rotates.

Whilst the spherical radial distribution of electronic charge about the nucleus cannot be described in terms of the Bessel function it is interesting to observe that the extensive degeneracy of the orbitals is removed in the presence of strong electric or magnetic fields. The initially spherical charge distribution must evidently distort to adopt an ellipsoidal symmetry, and the degenerate energy levels split up into a number of closely spaced non-degenerate levels. Spectral lines, arising from transitions to and from the initial levels are now accompanied by fine structure in the spectrum subject to certain selection rules regarding which transitions are allowed or forbidden. These are the Stark (electric field) and Zeeman (magnetic field) effects and the removal of degeneracy from the atomic stand-point is discussed in Section 4.10.

3.4 Fraunhofer Diffraction at a Circular Aperture

We shall consider two cases of the interaction of electromagnetic fields and matter, both of which are problems of cylindrical symmetry, and whose field distributions are described in terms of the Bessel functions. The problem with which we shall be concerned here involves the distribution of diffracted intensity on a screen placed a large distance from a circular aperture, that is, a large distance relative to the radius a of the aperture. This is, of course, the condition for Fraunhofer diffraction: Fresnel diffraction will occur when the screen is much closer to the diffracting aperture.

It is quite clear that the diffraction pattern is going to be rotationally symmetric about the z-axis, and we shall investigate the diffracted amplitude at a point P on the screen. Without any loss of generality, it will simplify the geometry if we consider P to lie on the yz plane, as shown in Figure 3.4.1, so that the cartesian

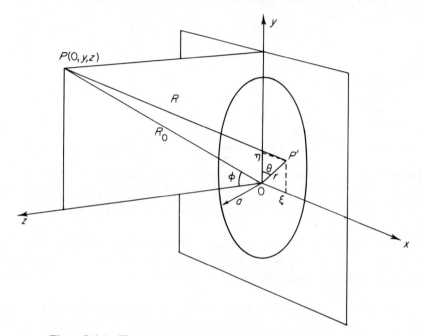

Figure 3.4.1 The geometry for the diffraction by a circular aperture

coordinate of P is $(0, y, z)$. We consider a point P' in the circular aperture at (ξ, η) and establish a relation between the separation PP' and the parameters of the configuration. In the cartesian frame we see that

$$R^2 = (\xi^2 + (y - \eta)^2 + z^2)$$

and

$$R_0^2 = y^2 + z^2$$

so that

$$R = (R_0^2 + \eta^2 + \xi^2 - 2\eta y)$$

Since

$$y = R_0 \sin \phi, \quad \xi = r \sin \theta, \quad \eta = r \cos \theta$$

then in polar coordinates

$$R = R_0 \left\{ 1 + \frac{r^2 \sin^2 \theta}{R_0^2} + \frac{r^2 \cos^2 \theta}{R_0^2} - \frac{2r}{R_0} \cos \theta \sin \phi \right\}^{1/2} \tag{3.4.1}$$

Further, the Fraunhofer condition sets $R_0 \gg r$, and so

$$R \simeq R_0 - r \cos \theta \sin \phi \qquad (3.4.2)$$

Now the amplitude of the diffracted field at P due to the element at P' will be

$$
\begin{aligned}
\delta A_P &= A \exp (ikR) r \, d\theta \, dr \\
&= A' \exp (-ikr \cos \theta \sin \phi) r \, d\theta \, dr
\end{aligned}
\qquad (3.4.3)
$$

where $A' = A \exp (ikR_0)$ and $k = 2\pi/\lambda$. At the point P we may set $\sin \phi = \alpha$, a constant. Total amplitude at P due to the whole aperture is then

$$A_P = A' \int_0^{2\pi} \int_0^a \exp (-ikr\alpha \cos \theta) r \, d\theta \, dr \qquad (3.4.4)$$

We could evaluate the double integral by expanding the exponential and integrating term by term, but it is much more convenient to use the integral definition of the zero-order Bessel function J_0 (see Appendix III):

$$\frac{1}{2\pi} \int_0^{2\pi} \exp (-ik\alpha r \cos \theta) \, d\theta = J_0(k\alpha r) \qquad (3.4.5)$$

Equation (3.4.4) then becomes

$$A_P = 2\pi A' \int_0^a J_0(k\alpha r) r \, dr \qquad (3.4.6)$$

and making the change of variable

$$x = k\alpha r$$

equation (3.4.6) becomes

$$A_P = \frac{2\pi A'}{k^2 \alpha^2} \int_0^{k\alpha a} x J_0(x) \, dx = \frac{2\pi A'}{k^2 \alpha^2} [x J_1(x)]_0^{k\alpha a} \qquad (3.4.7)$$

$$= 2\pi a^2 A' \frac{J_1(k\alpha a)}{k\alpha a} \qquad (3.4.8)$$

In this last step we have utilized the recurrence relation between Bessel functions (Appendix III)

$$x J_0 = \frac{d}{dx}(x J_1) \qquad (3.4.9)$$

The intensity at P follows directly from equation (3.4.8) as

$$I_P = (2\pi a^2)^2 A'^2 \left[\frac{J_1(k\alpha a)}{k\alpha a} \right]^2 \qquad (3.4.10)$$

The form of the first-order Bessel function of the first kind is shown in Figure 3.2.1(a), and its series expansion is

$$\frac{J_1(x)}{x} = \frac{1}{2} - \frac{x^2}{2!2^3} + \frac{x^4}{2!3!2^5} - \cdots \tag{3.4.11}$$

Thus when $\phi = 0$ (i.e. $x = 0$) we have a bright central maximum defining the so-called Airy disc. As x increases the amplitude passes through successive zeros, and the intensity distribution correspondingly passes through maxima and minima of intensity. In Figure 3.4.2 we show the Fraunhofer distribution of

Figure 3.4.2 The Fraunhofer distribution of diffracted intensity by a circular aperture

diffracted intensity by a circular aperture. The intensity distribution is seen to consist of circular fringes, and the minima are located at the zero points of $J_1(x)$. The circular locus of the nth dark fringe will be found at the radial distance x_n and the first few of these are tabulated in Table 3.4.1.

Table 3.4.1. Zeros of $J_1(x)$

x_0	0.7655π
x_1	1.2197π
x_2	1.6348π
x_3	2.0308π
x_4	2.4153π

We see from Table 3.4.1 that the radial loci of the minima are by no means harmonically related, and crowd in towards the edge of the pattern.

3.5 Diffusion in an Infinitely Long Cylinder

We shall now discuss the cylindrical solutions of the diffusion equation. As we observed in Section 2.10, these solutions differ essentially from those of the wave equation in that they are irreversible in time. This is a direct consequence of the appearance of the first time derivative in the diffusion equation, and this is not invariant to the transformation $t \rightarrow (-t)$ corresponding to the reversal of the time flow. If the wave equation contains a damping term, however, that is, a term proportional to the first time derivative, then again the solutions become irreversible and describe essentially dissipative processes.

The axially symmetric $(\partial\psi/\partial\phi = 0)$ diffusion equation in cylindrical coordinates is

$$\frac{\partial^2 \psi}{\partial r^2} + \frac{1}{r}\frac{\partial \psi}{\partial r} = \frac{1}{D}\frac{\partial \psi}{\partial t} \tag{3.5.1}$$

and we shall assume that the field varies sinusoidally at the surface of the cylinder, which is of radius a,

$$\psi(a, t) = S_0\, e^{i\omega t} \tag{3.5.2}$$

We are therefore justified in looking for solutions of the form

$$\psi(r, t) = R(r)\, e^{i\omega t} \tag{3.5.3}$$

Substitution of this form of solution in equation (3.5.1) yields

$$r^2 \frac{d^2 R}{dr^2} + r\frac{dR}{dr} - \frac{i\omega r^2 R}{D} = 0 \tag{3.5.4}$$

which is seen to be Bessel's equation of zero order and imaginary argument, having the solution

$$R(r) = A J_0(kr)$$

where

$$k\sqrt{-\frac{i\omega}{D}} = i\sqrt{i}\sqrt{\frac{\omega}{D}} \tag{3.5.5}$$

or in terms of the hyperbolic Bessel function, $R(r) = A'I_0(-ikr)$. If we expand the zero-order Bessel function (See Appendix III), we find it is a series of alternating real and imaginary terms:

$$R(r) = A\left\{1 + i\frac{\omega}{D}\left(\frac{r}{2}\right)^2 - \frac{1}{(2!)^2}\left(\frac{\omega}{D}\right)^2\left(\frac{r}{2}\right)^4 - \frac{i}{(3!)^3}\left(\frac{\omega}{D}\right)^3\left(\frac{r}{2}\right)^6 + \cdots\right\} \tag{3.5.6}$$

The real and imaginary components may be separated as follows:

$$R(r) = A\left\{\text{ber}_0\sqrt{\frac{\omega}{D}}r + i\,\text{bei}_0\sqrt{\frac{\omega}{D}}r\right\} \tag{3.5.7}$$

where we have the relations

$$J_n(xi\sqrt{i}) = i^n I_n(x\sqrt{i}) = \text{ber}_n\, x + i\,\text{bei}_n\, x$$

$$\text{Re}\, J_n(xi\sqrt{i}) = \text{ber}_n\, x$$

$$\text{Im}\,[J_n(xi\sqrt{i})] = \text{bei}_n\, x$$

Using these functions we may write our solution

$$\psi(r, t) = A\left\{\text{ber}_0\sqrt{\frac{\omega}{D}}r + i\,\text{bei}_0\sqrt{\frac{\omega}{D}}r\right\} e^{i\omega t}$$

$$= A\left\{\text{ber}_0^2\sqrt{\frac{\omega}{D}}r + \text{bei}_0^2\sqrt{\frac{\omega}{D}}r\right\}^{\frac{1}{2}} e^{i(\omega t + \phi)},\tag{3.5.8}$$

where

$$\tan\phi = \text{bei}\sqrt{\frac{\omega}{D}}r \Big/ \text{ber}\sqrt{\frac{\omega}{D}}r\tag{3.5.9}$$

Imposing the boundary condition (3.5.2) at $t = 0, r = a$ we have

$$A = S_0\left\{\text{ber}_0^2\sqrt{\frac{\omega}{D}}a + \text{bei}_0^2\sqrt{\frac{\omega}{D}}a\right\}^{-\frac{1}{2}} e^{-i\phi(a)}.\tag{3.5.10}$$

The envelope of this function as a function of the radius is shown in Figure 3.5.1, together with the variation of phase. The general form is similar to the solution of the analogous problem of the semi-infinite slab, although the precise functional form of the response is different. Nonetheless, both represent damped waves propagating inwards.

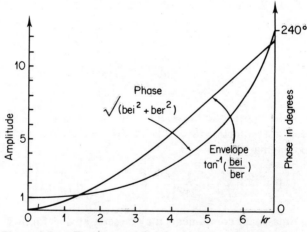

Figure 3.5.1 The ber and bei functions representing the radial variation of amplitude and phase of thermal waves in a cylindrical rod

3.6 Propagation of an Electromagnetic Wave in a Cylindrical Conductor

In a medium of finite conductivity σ, the Maxwell field relations between the quantities $\mathbf{E}, \mathbf{D}, \mathbf{B}, \mathbf{H}$, become

$$\text{div } \mathbf{D} = 0 \tag{3.6.1a}$$

$$\text{div } \mathbf{B} = 0 \tag{3.6.1b}$$

$$\text{curl } \mathbf{E} = -\partial \mathbf{B}/\partial t \tag{3.6.1c}$$

$$\text{curl } \mathbf{H} = \partial \mathbf{D}/\partial t + \sigma \mathbf{E} \tag{3.6.1d}$$

where, in equation (3.6.1d) the additional term $\sigma \mathbf{E}$ accounts for the development of conduction currents in the medium. Taking the curl of equation (3.6.1c) we obtain

$$\text{curl curl } \mathbf{E} = -\mu \frac{\partial}{\partial t} \text{curl } \mathbf{H}$$

i.e.

$$\text{grad div } \mathbf{E} - \nabla^2 \mathbf{E} = -\mu\varepsilon \frac{\partial^2 \mathbf{E}}{\partial t^2} - \mu\sigma \frac{\partial \mathbf{E}}{\partial t} \tag{3.6.2}$$

But since $\text{div } \mathbf{D} = \varepsilon \text{ div } \mathbf{E} = 0$, we have from equation (3.6.2) the general wave equation

$$\nabla^2 \mathbf{E} = \mu\varepsilon \frac{\partial^2 \mathbf{E}}{\partial t^2} + \mu\sigma \frac{\partial \mathbf{E}}{\partial t} \tag{3.6.3}$$

An exactly similar equation holds for $\mathbf{H}$. A notable feature of this differential equation is that it represents a damped wave motion due to the dissipative term $\mu\sigma(\partial \mathbf{E}/\partial t)$. In a dielectric, for which we have $\sigma = 0$, an unattenuated wave propagates through the medium, whereas in a metal we may expect the waveform to be very rapidly damped spatially. Indeed, in the case of metallic conduction the second derivative $(\partial^2 \mathbf{E}/\partial t^2)$ may be entirely neglected in comparison with the dissipative term, and writing $\mathbf{j}$ for $\sigma \mathbf{E}$ in equation (3.6.3) we have

$$\nabla^2 \mathbf{j} = \mu\sigma \left(\frac{\partial \mathbf{j}}{\partial t} \right) = i\omega\mu\sigma \mathbf{j} \tag{3.6.4}$$

where we have assumed a time dependence of the form $\exp(i\omega t)$. Expressing ∇^2 in cylindrical coordinates we obtain

$$\mathbf{j}(r, t) = R(r)T(t)$$

$$r^2 \frac{d^2 R}{dr^2} + r \frac{dR}{dr} - i\omega\sigma\mu r^2 R = 0 \tag{3.6.5}$$

which is seen, by comparison with equation (3.5.4), to be the Bessel equation of zero order:

$$R(r) = AJ_0(ks) \tag{3.6.6}$$

where $k = \sqrt{(-i\omega\sigma\mu)} = i\sqrt{i}\sqrt{(\omega\sigma\mu)}$.

Thus, as in the case of diffusion in an infinitely long cylinder (Section 3.5), the complex solution may be written

$$R(r) = A\{\text{ber}_0 \sqrt{(\omega\sigma\mu)}r + i\,\text{bei}_0 \sqrt{(\omega\sigma\mu)}r\}$$

The real envelope to this function, $A\{\text{ber}_0^2 \sqrt{(\omega\sigma\mu)}r + \text{bei}_0^2 \sqrt{(\omega\sigma\mu)}r\}$ is shown in Figure 3.5.1, as is the radial variation of phase within the conductor, $\phi = \tan^{-1}(\text{bei}_0 \sqrt{(\omega\sigma\mu)}r/\text{ber}_0 \sqrt{(\omega\sigma\mu)}r)$. The existence of a large alternating current vector $\mathbf{j}(r, t)$ at the surface of the conductor is known as the *skin effect*. As either ω or σ decreases so the envelope relaxes and at low frequencies, or under d.c. conditions, the current vector is uniformly distributed over the cross-section of the conductor.

The current distribution

$$A\{\text{ber}_0^2 \sqrt{(\omega\sigma\mu)}r + \text{bei}_0^2 \sqrt{(\omega\sigma\mu)}r\}^{\frac{1}{2}}$$

$$\sim A\left\{1 - \frac{(\omega\sigma\mu)^2}{(2!)^2}\left(\frac{r}{2}\right)^4 + \frac{(\omega\sigma\mu)^4}{(2!)^4}\left(\frac{r}{2}\right)^8 - \cdots\right\} \tag{3.6.7}$$

i.e.

$$\sim A\exp\left\{-\frac{\omega\sigma\mu}{2!}\left(\frac{r}{2}\right)^2\right\}$$

So that the distance over which the amplitude falls to a value of e^{-1} times its initial value is

$$\delta = \frac{1}{\sqrt{(\frac{1}{2}\omega\sigma\mu)}} \tag{3.6.8}$$

where δ is termed the *skin depth*. Thus the magnitude of the skin depth decreases with the inverse half-power of the frequency, permeability and conductivity of a metal. Values of the skin depth of copper at three different frequencies are given in Table 3.6.1.

Table 3.6.1. Skin depth in copper

Frequency (Mc/s)	δ(cm)
1·0	$6\cdot6 \times 10^{-3}$
10^4	$6\cdot6 \times 10^{-5}$
6×10^8	$2\cdot7 \times 10^{-7}$

For conductors of high permeability and conductivity such as iron, these skin effects may be apparent even at frequencies as low as 1 cycle/sec. We may qualitatively demonstrate the phase–amplitude relationships by a vector diagram shown in Figure 3.6.1.

It may be quite easily shown that the resistance of the cylindrical conductor is approximately that of a thin tube of the same radius, but of thickness δ carrying

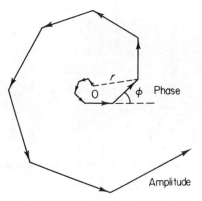

Figure 3.6.1 Amplitude–phase diagram for the current vector $\mathbf{j}(r)$

a uniformly distributed current. The a.c. resistivity of the conductor is therefore very much greater than its d.c. value (Figure 3.6.2).

At very high frequencies the skin depth δ becomes less than the electronic mean free path within the metal and the high frequency resistivity is then greater than that calculated on the basis of the classical formula (3.6.8). Under these circumstances the resistivity is governed rather by the time spent within the skin depth than the actual time τ between collisions. This is known as the *anomalous skin effect*.

As we have just seen, solutions to Maxwell's equation at the boundary between a conductor and free space will have appreciable amplitude only

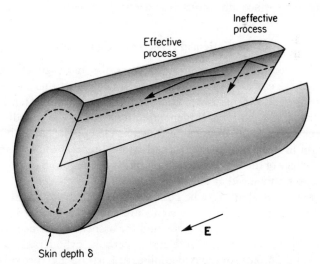

Figure 3.6.2 Skin effect showing contribution of surface electron transport to the conduction process. (From Gordon S. Kino and John Shaw, 'Acoustic surface waves'. Copyright © 1972 by Scientific American, Inc. All rights reserved)

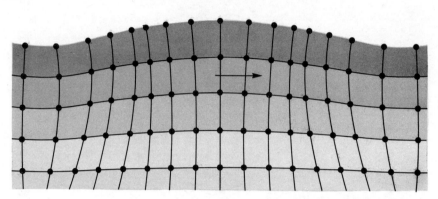

Figure 3.6.3 The Rayleigh or acoustic surface wave has both longitudinal and transverse components and develops specifically at the boundary between two media

within short distances of the boundary. Two associated physical phenomena are the existence of a 'ground wave' in radio transmission, and elastic *Rayleigh waves* which develop at the boundary between two media. As we see from Figure 3.6.3 these surface waves have both longitudinal and transverse components, and are not restricted to the description of the free solid surface. Indeed, such two-dimensional 'Debye' waves can be used to describe the high-frequency behaviour of the free liquid surface, and as such admit to the possibility of calculating the free energy (and hence surface tension) of such surface vibrational modes. In seismic work surface waves of a transverse type occur when a second medium overlies the elastic medium. Then *two* boundaries are involved and the surface vibrations are termed *Love waves*.

3.7 Elliptic and Hyperbolic Coordinates

The ellipsoidal coordinate system represents the most general coordinate frame in which we can attempt solutions of the field equations discussed in this book. By means of various distortions, compressions and extensions of the ellipsoidal frame it is possible to generate all eleven of the coordinate frames generally encountered in physical problems. Of course, separability of the field equation in these more exotic coordinate systems is another matter, and the full machinery of the Stäckel determinant may be required before any sort of separation can be achieved at all.

Here, however, we shall only be concerned with the cylindrical elliptic and hyperbolic coordinates and these systems are shown in Figure 3.7.1. The foci A, B are located on the x-axis at $\pm a/2$, and the associated surfaces constitute a family of *orthogonal* confocal elliptic and hyperbolic cylinders. It is important to notice that the surfaces of the two families are orthogonal (provided they are confocal) since one set is related to the gradient of the other, and so one family will generally represent equipotential surfaces whilst the other will represent the associated stream function.

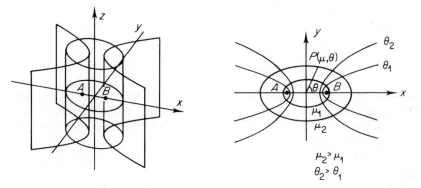

Figure 3.7.1 The cylindrical elliptic and hyperbolic coordinates

If we set

$$x = \tfrac{1}{2}a \cosh \mu \cos \theta, \quad y = \tfrac{1}{2}a \sinh \mu \sin \theta$$

then curves of constant μ are ellipses, and curves with constant θ are hyperbolas, all with foci at $\pm a/2$. μ may be understood as the new radial parameter and ranges $0 \leqslant \mu \leqslant \infty$. Whereas in polar coordinates $r = $ constant described a circle, here $\mu = $ constant describes an ellipse. The θ-coordinate retains the same significance as in the polar case, having the range $0 \leqslant \theta \leqslant 2\pi$. The confocal elliptic cylinders thus described have major axis $a \cdot \cosh \mu$, and minor axis $a \cdot \sinh \mu$.

A particularly important situation is represented by the condition $\mu = 0$ when the ellipse degenerates into a flat strip of width a centred along the z-axis. Correspondingly, the condition $\theta = 0$ reduces the hyperbolic surface to the entire x, z plane *except* for the central strip of width a, again centred on the z-axis.

In this coordinate system the gradient operator has the form

$$\nabla \equiv \frac{2/a}{\sqrt{(\cosh^2 \mu - \cos^2 \theta)}} \left\{ \hat{\boldsymbol{\mu}} \frac{\partial}{\partial \mu} + \hat{\boldsymbol{\theta}} \frac{\partial}{\partial \theta} \right\} \tag{3.7.1}$$

where $\hat{\boldsymbol{\mu}}$ and $\hat{\boldsymbol{\theta}}$ are the unit vectors, and the Laplacian becomes

$$\nabla^2 \equiv \frac{4}{\cosh^2 \mu - \cos^2 \theta} \left\{ \frac{\partial}{\partial \mu^2} + \frac{\partial^2}{\partial \theta^2} \right\} \tag{3.7.2}$$

Laplace's equation $\nabla^2 \psi(\mu, \theta) = 0$ is particularly simple in these coordinates, reducing to

$$\frac{\partial^2 \psi}{\partial \mu^2} + \frac{\partial^2 \psi}{\partial \theta^2} = 0 \tag{3.7.3}$$

Assuming a solution of the form $\psi(\mu, \theta) = R(\mu)\Theta(\theta)$ equation (3.7.3) separates as follows

$$\frac{1}{R}\frac{d^2 R}{d\mu^2} = m^2; \quad \frac{1}{\Theta}\frac{d^2 \Theta}{d\theta^2} = -m^2 \tag{3.7.4}$$

The radial solutions are of the form $e^{\pm m\mu}$ or $\sinh m\mu$, $\cosh m\mu$ whilst the angular solutions are of the form $\sin m\theta$, $\cos m\theta$, for $|m| > 0$. It therefore follows that for the trigonometric solutions m must be an integer to ensure continuity of the angular solution at $\theta = 0, 2\pi$.

We may now consider a few simple examples of the application of Laplace's equation in its elliptical form.

The flow of an incompressible, irrotational non-viscous fluid along the y-axis through a slit of width a represents a simple hydrodynamic example of the solution of Laplace's equation for the velocity potential ψ in the fluid. To generate the slit from the elliptic coordinates we set $\theta = 0$, and consequently the Laplace equation reduces directly to

$$\frac{d^2 \psi(\mu)}{d\mu^2} = 0 \tag{3.7.5}$$

which has the solution $\psi(\mu) = A\mu$, where A is an arbitrary constant. The velocity equipotentials are evidently a family of confocal elliptic cylinders, and the associated stream functions are given as the orthogonal sheets shown in Figure 3.7.2. The fluid velocity across the slit, that is, at $\mu = 0$ is, from equation (3.7.1),

$$v = (\text{grad } \psi)_{\mu = 0} = \frac{A\mathbf{j}}{\sqrt{[(a/2)^2 - x^2]}} \tag{3.7.6}$$

where $\mathbf{j}$ is the unit flow vector when $\mu = 0$.

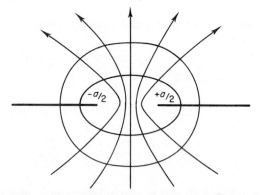

Figure 3.7.2 The flow lines and velocity equipotentials for an incompressible, irrotational non-viscous fluid flowing through an infinitely long slit of width a placed perpendicular to the normal flow

As an electrostatic example of the use of Laplace's equation we consider the potential distribution within the confocal system of a grounded elliptical cylinder $\mu = \mu_0$, and a flat strip of width a ($\mu = 0$) held at a potential V, Figure 3.7.3.

Equipotentials

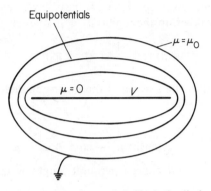

Figure 3.7.3 Grounded elliptical cylinder $\mu = \mu_0$ and confocal flat strip $\mu = 0$, held at potential V

The potential distribution within the system will be elliptically symmetric (cf. circularly symmetric), and will therefore be a function of μ only. Laplace's equation again reduces to

$$\frac{d^2\psi(\mu)}{d\mu^2} = 0; \quad \psi(\mu) = A\mu + c \tag{3.7.7}$$

where A and c are arbitrary constants. The boundary conditions $\psi(0) = V$, and $\psi(\mu_0) = 0$ enable us to determine the constants to be

$$A = -V/\mu_0, \quad c = V \tag{3.7.8}$$

so that

$$\psi(\mu) = V\left(1 - \frac{\mu}{\mu_0}\right), \quad 0 \leqslant \mu \leqslant \mu_0 \tag{3.7.9}$$

The field at the strip $\mu = 0$ is from equations (3.7.1) and (3.7.9)

$$\mathbf{E}_{\mu=0} = (-\nabla\psi)_{\mu=0} = -\frac{2/a}{\sqrt{(1 - \cos^2\theta)}}\left(\frac{\partial\psi}{\partial\mu}\right)_{\mu=0}$$

$$= \frac{2V}{\mu_0 a|\sin\theta|} \tag{3.7.10}$$

and hence the charge density $\sigma(\theta) = \varepsilon_0 E = 2\varepsilon_0 V/(\mu_0 a|\sin\theta|)$. The total charge on the strip follows as the integral over the surface of the strip

$$Q = 4\int_0^{\pi/2} \frac{2\varepsilon_0 V}{\mu_0 a \sin\theta} \frac{a}{2}\sin\theta \, d\theta = \frac{2\pi V\varepsilon_0}{\mu_0} \tag{3.7.11}$$

and hence the capacitance per unit length of the system

$$C = Q/V = 2\pi\varepsilon_0/\mu_0 \tag{3.7.12}$$

A more general case concerns two confocal elliptic cylinders μ_1 and μ_2 ($\mu_1 < \mu_2$) held at potentials V and zero respectively. It may be very easily shown that the distribution of potential within the region $\mu_1 \leqslant \mu \leqslant \mu_2$ is

$$\psi(\mu) = \frac{V}{\mu_1 - \mu_2}(\mu - \mu_2) \tag{3.7.13}$$

whilst the normal field at the surface of the inner conductor μ_1, is now, from equation (3.7.1),

$$\mathbf{E}(\mu_1) = \frac{2/a}{\sqrt{(\cosh^2 \mu_1 - \cos^2 \theta)}} \frac{V}{\mu_2 - \mu_1} \tag{3.7.14}$$

and the corresponding total charge per unit length is given as the elliptic integral

$$Q = \frac{4V\varepsilon_0}{\mu_2 - \mu_1} \int_0^{2\pi} \frac{2}{a\sqrt{(\cosh^2 \mu_1 - \cos^2 \theta)}} \frac{a}{2}\sqrt{(\cosh^2 \mu_1 - \cos^2 \theta)} \, d\theta$$
$$= 2\pi\varepsilon_0 V/(\mu_2 - \mu_1) \tag{3.7.15}$$

and the capacitance

$$C = \frac{2\pi\varepsilon_0}{\mu_2 - \mu_1} \tag{3.7.16}$$

which is seen to reduce to equation (3.7.12) as $\mu_1 \to 0, \mu_2 \to \mu_0$.

3.8 Hydrodynamic Irrotational Flow Past an Elliptical Aerofoil

We now consider the hydrodynamic irrotational flow of an incompressible fluid past an elliptical profile μ_0 at an angle γ to the general fluid velocity. The (x, y) coordinate frame is set up on the major axis as shown in Figure 3.8.1.

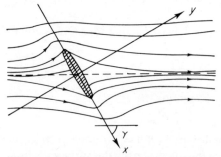

Figure 3.8.1 Hydrodynamic irrotational flow about an elliptic aerofoil. γ is the angle of attack, or angle to the general flow

Clearly, the velocity potential for uniform flow with velocity in the y-direction is

$$\psi_0(x \cos \gamma + y \sin \gamma) = \tfrac{1}{2}av_0(\cosh \mu \cos \theta \cos \gamma + \sinh \mu \sin \theta \sin \gamma) \quad (3.8.1)$$

whilst in the vicinity of the ellipse the velocity potential will be given as the general solution to the Laplace equation, (3.7.4), which is finite as $\mu \to \infty$:

$$\psi(\mu, \theta) = \sum_{m=0}^{\infty} (a_m \sin m\theta + b_m \cos m\theta) e^{-m\mu} \quad (3.8.2)$$

where m must be an integer for continuity in angle. From the asymptotic form of the velocity potential (3.8.1) it is evident that we are interested in solutions for which $m = 1$ and consequently we adopt the solution

$$\psi(\mu, \theta) = (a_1 \sin \theta + b_1 \cos \theta) e^{-\mu}$$
$$+ \frac{av_0}{2}[\cosh \mu \cos \theta \cos \gamma + \sinh \mu \sin \theta \sin \gamma] \quad (3.8.3)$$

Since the fluid cannot flow into the cylinder, that is, the normal component of the velocity is zero at the surface of the cylinder, we have the Neumann condition $(\partial \psi / \partial \mu)_{\mu = \mu_0} = 0$, and this enables us to determine the arbitrary coefficients a_1, b_1 as follows:

$$\left(\frac{\partial \psi}{\partial \mu}\right) = -a_1 \sin \theta\, e^{-\mu} - b_1 \cos \theta\, e^{-\mu}$$
$$+ \frac{av_0}{2}[\sinh \mu \cos \theta \cos \gamma + \cosh \mu \sin \theta \sin \gamma]$$

Comparing coefficients of $\sin \theta$, $\cos \theta$ at $\mu = \mu_0$, that is, when $(\partial \psi / \partial \mu) = 0$, we obtain

$$\left. \begin{aligned} a_1 &= \frac{av_0}{2} \sin \gamma\, e^{\mu_0} \cosh \mu_0 \\[2mm] b_1 &= \frac{av_0}{2} \cos \gamma\, e^{\mu_0} \sinh \mu_0 \end{aligned} \right\} \quad (3.8.4)$$

so that the final expression for the velocity potential becomes

$$\psi(\mu, \theta) = \frac{av_0}{2}[\cos \theta \cos \gamma(\cosh \mu + e^{\mu_0 - \mu} \sinh \mu_0)$$
$$+ \sin \theta \sin \gamma(\sinh \mu + e^{\mu_0 - \mu} \cosh \mu_0)] \quad (3.8.5)$$

This, of course, reduces to the specific case of a strip when we let $\mu_0 \to 0$.

The velocity of the fluid past the general elliptical surface is the gradient evaluated at μ_0:

$$v = |\text{grad } \psi|_{\mu_0} = \frac{2/a}{\sqrt{(\cosh^2 \mu_0 - \cos^2 \theta)}}\left(\frac{2\psi}{\partial \theta}\right)_{\mu_0}$$

$$= \frac{v_0\, e^{\mu_0} \sin(\gamma - \theta)}{\sqrt{(\cosh^2 \mu_0 - \cos^2 \theta)}} \tag{3.8.6}$$

We may directly investigate the pressure distribution over the surface of the cylinder, for, from Bernoulli's equation the pressure is related to the velocity distribution as

$$p(\mu_0, \theta) = p_0 + \tfrac{1}{2}\rho[v_0^2 - v^2(\mu_0, \theta)] \tag{3.8.7}$$

where p_0 is the hydrostatic pressure at a point a long way from the aerofoil (and in a gravitational field will be a function of depth) and ρ is the fluid density. Inserting equation (3.8.6) into equation (3.8.7) we obtain the general pressure distribution over the cylindrical surface $\mu = \mu_0$ as

$$p(\mu_0, \theta) = p_0 + \tfrac{1}{2}\rho v_0^2\left\{1 - \frac{e^{2\mu_0} \sin^2(\gamma - \theta)}{\cosh^2 \mu_0 - \cos^2 \theta}\right\} \tag{3.8.8}$$

We see from this expression that for given μ_0, γ the pressure is smallest in the regions of greatest curvature, that is, at $\theta = \pi, 2\pi$—the leading and trailing edges with respect to the flow. Indeed for small values of the parameter μ_0 corresponding to a very flattened elliptical profile the pressure may even become negative, i.e. as $\mu_0 \to 0$. The obvious limiting case is for $\mu_0 = 0$, representing a flat strip of width a oriented at γ to the general flow. Under these circumstances Bernoulli's equation (3.8.8) reduces to

$$p(0, \theta) = p_0 + \tfrac{1}{2}\rho v_0^2\left\{1 - \frac{\sin^2(\gamma - \theta)}{1 - \cos^2 \theta}\right\} \tag{3.8.9}$$

this representing the pressure distribution over the flat strip. We see that at the trailing edge ($\theta = 0$) the pressure tends to $-\infty$, as indeed it does at the leading edge. Physically this means that the fluid will not remain in contact with the aerofoil surface and steady irrotational flow is no longer possible. A less acute situation pertains for profiles having $\mu_0 > 0$; regions of negative pressure will nevertheless develop but will not be negatively infinite. For a given fluid velocity and orientation of the aerofoil with respect to the flow we may easily establish the criterion

$$\mu_0 \geqslant \cosh^{-1}\left\{1 + \frac{\rho v_0^2 \sin^2 \gamma}{\rho v_0^2 + 2p_0}\right\}^{\frac{1}{2}} \tag{3.8.10}$$

for steady irrotational hydrodynamic flow over the surface. We see from this that irrotational flow around a strip aerofoil can only occur if the angle of attack γ is zero. Irrotational flow will continue as the orientation is increased provided the profile μ_0 varies in accordance with the inequality (3.8.10). We should notice that the pressure distribution over the aerofoil for irrotational non-viscous hydrodynamic flow is perfectly symmetrical. The significance of this is that there is no net force on the aerofoil. We shall see shortly that the incorporation of circulation into the velocity potential will establish *lift*.

The physical consequence of such a pressure distribution (3.8.9) over a flat aerofoil is that irrotational flow of the type described in equation (3.8.8) with $\mu_0 = 0$ is impossible: the streamlines are shown in Figure 3.8.2(a). Physically, vortices slide off the trailing edge setting up a counterrotation around the aerofoil thereby lowering the fluid velocity over the sharp edge. The effect on the flow of including a circulation is shown in Figure 3.8.2(b). In writing down the

(a)   (b)

Figure 3.8.2   (a) The symmetric irrotational flow about a flat aerofoil at angle γ to the general flow. (b) The flow about a flat aerofoil incorporating rotation. The flow is no longer symmetric, and there is a net 'lift' on the aerofoil

solutions (3.8.2), (3.8.3) to Laplace's equation for the velocity potential we neglected the possibility of multivalued solutions of the form $A\theta$. Physically this represents the development of a vortex filament, and has been discussed earlier in this chapter (Section 3.1). The associated velocity is, of course, $2A/a\sqrt{(\cosh^2 \mu - \cos^2 \theta)}$ at the point (μ, θ) and represents a *rotational* component to the fluid flow at the point μ in the direction θ. If we now incorporate this term in the expression for the velocity potential (3.8.5), and evaluate the gradient at the point $\mu = 0, \theta = 0$ for $\mu_0 = 0$ and set it equal to zero, then the coefficient $A = -\sin \gamma$. The velocity potential in the present case becomes

$$\psi(\mu, \theta) = \tfrac{1}{2}av_0\{\cos \gamma \cos \theta \cosh \mu + \sin \gamma \sin \theta \cosh \mu - \theta \sin \gamma\} \quad (3.8.11)$$

and

$$p(0, \theta) = p_0 + \tfrac{1}{2}\rho v_0^2 \left\{\sin^2 \gamma \left[1 - \tan^2 \left(\frac{\theta}{2}\right)\right] - \sin 2\gamma \tan \left(\frac{\theta}{2}\right)\right\} \quad (3.8.12)$$

The pressure distribution is now markedly different to that of the irrotational case: the $\tan (\theta/2)$ term is responsible for a greater pressure being exerted in the region $\pi < \theta < 2\pi$, i.e. on the underside, than on the topside. The physical consequence of the net circulation is, of course, *lift*. This problem is considered in some detail by Morse and Feshbach. We have made no comment on what happens at the leading edge ($\theta = \pi$) where for irrotational flow vortices are again generated. It may happen that the vortices are trapped on the leading edge by the flow creating a 'dead space', effectively reducing the curvature and making subsequent generation of vortices in this region less likely. In practice, of course, the leading edges of aerofoils are rounded and the situation does not develop.

The fluid velocity $\nabla\psi$ appears to have a rotational component $-v_0 \sin \gamma / \sqrt{\cosh^2 \mu - \cos^2 \theta}$ even at very large distances from the aerofoil, and this is difficult to understand. Morse and Feshbach suggest that beyond a certain distance not too far from the aerofoil there may be a net cancellation between the circulation and the vortices which are thrown off the trailing edge.

3.9 Potential Distribution Around an Elliptic Cylinder

The solution of Laplace's equation for the distribution of electrostatic potential around the elliptic cylinder μ_0 may conveniently be divided into the regions $\mu > \mu_0, \mu < \mu_0$ relating to the internal and external distributions. The cylinder is assumed to have an arbitrary surface potential $\psi_0(\mu_0, \theta)$ and we require the field to tend to zero as $\mu \to \infty$. Assuming a potential distribution of the form $\psi(\mu, \theta) = R(\mu)\Theta(\theta)$ the Laplace equation yields the two separated equations

$$\frac{1}{R}\frac{d^2R}{d\mu^2} = m^2; \quad \frac{1}{\Theta}\frac{d^2\Theta}{d\theta^2} = -m^2$$

where m is an integer. The general solution for $\mu > \mu_0$ may then be written

$$\psi(\mu, \theta) = \sum_{m=0}^{\infty} (A_m \sin m\theta + B_m \cos m\theta) e^{-m\mu}$$

$$= B_0 + \sum_{m=1}^{\infty} (A_m \sin m\theta + B_m \cos m\theta) e^{-m\mu} (\mu > \mu_0)$$

(3.9.1)

for which the field tends to zero as $\mu \to \infty$.

The general solution inside the elliptic cylinder $\mu < \mu_0$ is

$$\psi(\mu, \theta) = \sum_{m=0}^{\infty} C_m \sin m\theta \sinh m\mu + \sum_{m=0}^{\infty} D_m \cos m\theta \cosh m\mu$$

$$= D_0 + \sum_{m=1}^{\infty} C_m \sin m\theta \sinh m\mu + \sum_{m=1}^{\infty} D_m \cos m\theta \cosh m\mu$$

(3.9.2)

where the hyperbolic solutions have been adopted. Had the exponential solutions been retained a cusp in the potential would have developed at $\mu = 0$, and we are seeking continuous solutions. We may easily determine the arbitrary coefficients A_m, B_m, etc., by exploiting the orthogonality of the trigonometric functions. Thus over the surface μ_0 we have, from equation (3.9.1), multiplying throughout by $\sin m\beta$ and integrating:

$$\frac{1}{\pi}\int_0^{2\pi} \psi(\mu_0, \beta) \sin m\beta \, e^{m\mu_0} \, d\beta = A_m$$

(3.9.3)

and by $\cos m\beta$

$$\frac{1}{\pi} \int_0^{2\pi} \psi_0(\mu_0, \beta) \cos m\beta \, e^{m\mu_0} \, d\beta = B_m$$

so that the general solution for $\mu > \mu_0$ becomes

$$\psi(\mu, \theta) = \sum_{m=0}^{\infty} \left\{ \frac{1}{\pi} \int_0^{2\pi} \psi_0(\mu_0, \beta) \cos m(\theta - \beta) \, d\beta \right\} e^{m(\mu_0 - \mu)}$$

$$= \frac{1}{\pi} \int_0^{2\pi} \psi_0(\mu_0, \beta) \, d\beta + \sum_{m=1}^{\infty} \frac{1}{\pi} \int_0^{2\pi} \{\psi_0(\eta_0, \beta) \cos m(\theta - \beta) \, d\beta\} \, e^{m(\mu_0 - \mu)}$$

$$(3.9.4)$$

where the coefficient B_0 evidently represents the angle-averaged or net potential of the cylinder.

Similarly, in the region $\mu < \mu_0$ we have at the boundary

$$\frac{1}{\pi} \int_0^{2\pi} \psi_0(\mu_0, \beta) \sin m\beta \, d\beta = C_m \sinh \mu_0$$

$$\frac{1}{\pi} \int_0^{2\pi} \psi_0(\mu_0, \beta) \cos m\beta \, d\beta = D_m \cosh \mu_0$$

yielding the general solution for $\mu < \mu_0$

$$\psi(\mu, \theta) = \sum_{m=0}^{\infty} \left\{ \frac{1}{\pi} \int_0^{2\pi} \psi_0(\mu_0, \beta) \sin m\beta \, d\beta \right\} \sin m\theta \frac{\sinh m\mu}{\sinh m\mu_0}$$

$$+ \sum_{m=0}^{\infty} \left\{ \frac{1}{\pi} \int_0^{2\pi} \psi_0(\mu_0, \beta) \cos m\beta \, d\beta \right\} \cos m\theta \frac{\cosh m\mu}{\cosh m\mu_0}$$

$$= \frac{1}{\pi} \int_0^{2\pi} \psi_0(\mu_0, \beta) \cos m\beta \, d\beta \qquad (3.9.5)$$

$$+ \sum_{m=1}^{\infty} \left\{ \frac{1}{\pi} \int_0^{2\pi} \psi_0(\mu_0, \beta) \sin m\beta \, d\beta \right\} \sin m\theta \frac{\sinh m\mu}{\sinh m\mu_0}$$

$$+ \sum_{m=1}^{\infty} \left\{ \frac{1}{\pi} \int_0^{2\pi} \psi_0(\mu_0, \beta) \cos m\beta \, d\beta \right\} \cos m\theta \frac{\cosh m\mu}{\cosh m\mu_0}$$

Evidently $B_0 = D_0$. Expressions (3.9.4) and (3.9.5) are seen to become identical at $\mu = \mu_0$ as they should. Moreover, the potential inside the elliptic cylinder remains continuous, and hence the field remains finite at $\mu = 0$.

The potential distribution about an elliptic cylinder μ_0 of dielectric constant ε immersed at an angle γ in a uniform electric field will now be considered. The potential associated with the uniform field will be

$$- E(x \cos \gamma + y \sin \gamma) = -\tfrac{1}{2} Ea(\cosh \mu \cos \theta \cos \gamma + \sinh \mu \sin \theta \sin \gamma)$$

$$(3.9.6)$$

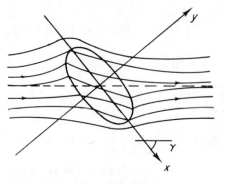

Figure 3.9.1 The distortion of the local electric field in the vicinity of a dielectric elliptic cylinder whose major axis is oriented at γ to the uniform field

with respect to the cartesian frame shown in Figure 3.9.1. The asymptotic form of the potential as $\mu \to \infty$ suggests a potential of the form

$$\psi_1(\mu, \theta)_{\mu > \mu_0} = (a \sin \theta + b \cos \theta) e^{\mu_0 - \mu} - E(x \cos \gamma + y \sin \gamma) \quad (3.9.7)$$

whilst inside the boundary

$$\psi_2(\mu, \theta)_{\mu < \mu_0} = c \sin \theta \frac{\sinh \mu}{\sinh \mu_0} + d \cos \theta \frac{\cosh \mu}{\cosh \mu_0} \quad (3.9.8)$$

The arbitrary constants a, b and c, d may be determined from the condition that $(\partial\psi_1/\partial\mu)_{\mu_0} = \varepsilon(\partial\psi_2/\partial\mu)_{\mu_0}$ and $(\partial\psi_1/\partial\theta)_{\mu_0} = (\partial\psi_2/\partial\theta)_{\mu_0}$. The resulting potential distributions are:

$$\psi_1(\mu, \theta)_{\mu > \mu_0} = -E(x \cos \gamma + y \sin \gamma)$$

$$+ \tfrac{1}{4}Ea(\varepsilon - 1) \sinh (2\mu_0) e^{\mu_0 - \mu} \left\{ \frac{\cos \theta \cos \gamma}{\cosh \mu_0 + \varepsilon \sinh \mu} \right.$$

$$\left. + \frac{\sin \theta \sin \gamma}{\varepsilon \cosh \mu_0 + \sinh \mu_0} \right\} \quad (3.9.9)$$

$$\psi_2(\mu, \theta)_{\mu < \mu_0} = -E e^{\mu_0} \left\{ \frac{x \cos \gamma}{\cosh \mu_0 + \varepsilon \sinh \mu_0} + \frac{y \sin \gamma}{\varepsilon \cosh \mu_0 + \sinh \mu_0} \right\} \quad (3.9.10)$$

The field inside the dielectric ellipse is uniform, but has a different magnitude and direction to the external field. Charges are induced on the surface of the cylinder by the external field and the internal field is therefore the vector combination of the external field **E** and the polarization field $-\mathbf{P}$. For a sphere these are parallel of course, but in the present case the field $-\mathbf{P}$ is at an angle $(90 - \gamma)$ to the external field. The resulting internal field $\mathbf{E}_{int}$ therefore differs both in magnitude and direction to the applied field (Figure 3.9.2).

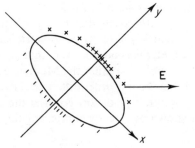
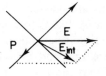

Figure 3.9.2 Since the field arising from the polarization charges induced on the surface of the cylinder is not exactly antiparallel to the applied field, the resultant internal field $\mathbf{E}_{int}$ generally differs in direction to the applied field

3.10 Elliptic Solutions of the Helmholtz Equation

So far we have only considered potential fields arising as solutions to Laplace's equation. In our consideration of normal mode solutions to the wave equation we should also now examine the Helmholtz equation in ellipitical coordinates (cf. equation (3.3.2))

$$\frac{\partial^2 \psi}{\partial \mu^2} + \frac{\partial^2 \psi}{\partial \theta^2} + \frac{1}{4}a^2 k^2 (\cosh^2 \mu - \cos^2 \theta)\psi = 0 \tag{3.10.1}$$

a is the separation of the foci and k^2 the usual separation constant arising in the Helmholtz equation. If we assume a separable solution to equation (3.10.1) of the form $\psi(\mu, \theta) = R(\mu)\Theta(\theta)$, then we immediately obtain

$$\frac{d^2 \Theta}{d\theta^2} + (b - h^2 \cos^2 \theta)\Theta = 0 \tag{3.10.2}$$

and

$$-\frac{d^2 R}{d\mu^2} + (b - h^2 \cosh^2 \mu)R = 0 \tag{3.10.3}$$

where $h = \frac{1}{2}ak = \pi a/\lambda$, and b is a separation constant. Both of these are, of course, Mathieu's equation which we have already encountered in our solution of the Schrödinger equation for a linear periodic potential (Section 2.9). Equation (3.10.3), however, is the Mathieu equation for an imaginary argument and represents the radial solution. We notice that as $h \to 0$ so we regain the trigonometric solutions for $\Theta(\theta)$, that is,

$$Se_m(h, \cos\theta) \to \cos m\theta; \quad So_m(h, \cos\theta) \to \sin m\theta$$

and correspondingly

$$b_m \to m^2$$

Unfortunately, the full complement of angular solutions is four: $So_{2m}(h, \cos\theta)$, $So_{2m+1}(h, \cos\theta)$, $Se_{2m}(h, \cos\theta)$ and $Se_{2m+1}(h, \cos\theta)$ depending whether the solutions are odd or even about the origin $\theta = 0$ (cf. trigonometric solutions $\sin\theta, \cos\theta$), and as to whether the parameter m is odd or even, that is, whether the function is periodic in π or 2π. This is not to say that the Mathieu equation has more independent solutions than other second-order differential equations, merely that the *range* of the angular solution (π or 2π) governs the form of solution. This is perhaps most clearly shown by directly comparing the various solutions (Table 3.10.1).

Table 3.10.1. Angular Mathieu functions

Even functions	$Se_{2m}(h, \cos\theta) = \sum\limits_{n=0}^{\infty} B_{2n}^e(h, 2m) \cos(2n\theta)$	of period π
	$Se_{2m+1}(h, \cos\theta) = \sum\limits_{n=0}^{\infty} B_{2n+1}^e(h, 2m+1) \cos((2n+1)\theta)$	of period 2π
Odd functions	$So_{2m}(h, \sin\theta) = \sum\limits_{n=0}^{\infty} B_{2n}^o(h, 2m) \sin(2n\theta)$	of period π
	$So_{2m+1}(h, \sin\theta) = \sum\limits_{n=0}^{\infty} B_{2n+1}^o(h, 2m+1) \sin((2n+1)\theta)$	of period 2π

Not surprisingly the angular eigenfunctions, So, Se, may be quite simply expressed in terms of the odd and even trigonometric functions $\cos n\theta$, $\sin n\theta$. The series expansions are summarized in Table 3.10.1. For given h, these constitute a complete set of orthogonal eigenfunctions.

Correspondingly, the radial solutions can be written in terms of the non-periodic argument $\cosh\mu$ as $So_{2m}(h, \cosh\mu)$, etc. but these functions may be directly, simply and equivalently expressed in terms of an expansion in Bessel functions when the solutions are termed Jo, Je, etc., which is perhaps more satisfactory physically. Table 3.10.2 summarizes the situation: again, the proliferation of solutions is a characteristic of the Mathieu equation. The radial and angular solutions associate in the appropriate way, that is, $Se_m(h, \cos\theta) [Je_m(h, \cosh\mu) + Ne(h, \cosh\mu)]$ and the corresponding combination $So_m(Jo_m + No_m)$. These solutions reduce in the limit $h \to 0$ to the trigonometric angular solution in product with the Bessel function. The example given above, for instance, reduces to $\cos m\theta[J_m(\cosh\mu) + N_m(\cosh\mu)]$ as indeed it should, for the condition $h \to 0$ is equivalent to the foci becoming coincident ($a \to 0$) in which case we should regain the circular solutions. The radial solutions to *interior* problems are generally restricted to the radial functions of the first kind Jo, Je since No, Ne become infinite at $\mu = 0$. An analogous situation holds of course for the Bessel functions of the second kind which are also usually employed only in *exterior* problems.

Table 3.10.2. Radial Mathieu functions (for values of B and b corresponding to the angular functions Se, So)

Of the first kind

Even functions

$$\mathrm{Je}_{2m}(h, \cosh \mu) = \sqrt{\frac{\pi}{2}} \sum_{n=0}^{\infty} (-1)^{n-m} B_{2n}^{e} J_{2n}(h, \cosh \mu)$$

$$\mathrm{Je}_{2m+1}(h, \cosh \mu) = \sqrt{\frac{\pi}{2}} \sum_{n=0}^{\infty} (-1)^{n-m} B_{2n+1}^{e} J_{2n+1}(h, \cosh \mu)$$

Odd functions

$$\mathrm{Jo}_{2m}(h, \cosh \mu) = \sqrt{\frac{\pi}{2}} \tan h\mu \sum_{n=0}^{\infty} (-1)^{n-m}(2n) B_{2n}^{o} J_{2n}(h, \cosh \mu)$$

$$\mathrm{Jo}_{2m+1}(h, \cosh \mu) = \sqrt{\frac{\pi}{2}} \tanh \mu \sum_{n=0}^{\infty} (-1)^{n-m}(2n+1) B_{2n+1}^{o} J_{2n+1}(h, \cosh \mu)$$

Of the second kind

Even functions

$$\mathrm{Ne}_{2m}(h, \cosh \mu) = \sqrt{\frac{\pi}{2}} \sum_{n=0}^{\infty} (-1)^{n-m} B_{2n}^{e} Y_{2n}(h, \cosh \mu)$$

$$\mathrm{Ne}_{2m+1}(h, \cosh \mu) = \sqrt{\frac{\pi}{2}} \sum_{n=0}^{\infty} (-1)^{n-m} B_{2n+1}^{e} Y_{2n+1}(h, \cosh \mu)$$

Odd functions

$$\mathrm{No}_{2m}(h, \cosh \mu) = \sqrt{\frac{\pi}{2}} \tanh \mu \sum_{n=0}^{\infty} (-1)^{n-m}(2n) B_{2n}^{o} Y_{2n}(h, \cosh \mu)$$

$$\mathrm{No}_{2m+1}(h, \cosh \mu) = \sqrt{\frac{\pi}{2}} \tanh \mu \sum_{n=0}^{\infty} (-1)^{n-m}(2n+1) B_{2n+1}^{o} Y_{2n+1}(h, \cosh \mu)$$

It is now a comparatively easy matter to determine the normal modes of vibration of an elliptic membrane μ_0. The solutions are of the form $\mathrm{Se}_m(h, \cos \theta) \, \mathrm{Je}_m(h, \cosh \mu)$ or $\mathrm{So}_m(h, \cos \theta) \, \mathrm{Jo}_m(h, \cosh \mu)$ where the eigenvalues are determined from the tabulated functions as the roots of $\mathrm{Je}_m(h, \cosh \mu_0) = 0$, just as for the Bessel functions.

When we come to discuss the solution of the Helmholtz equation in spheroidal coordinates (which are obtained by rotations about the major and minor elliptic axes) we shall not be surprised to find that the angular solutions are very conveniently expressed as a series expansion in the surface spherical harmonics. Moreover, the radial solutions are expressed as sums over the spherical Bessel functions of the first and second kinds.

Problems

1 Show that the internal and external distributions of potential $\psi(r)$ for a cylinder of radius a are of the form

$$\psi(r) = A \exp(-kr^2) + B \quad r < a$$
$$= -C \ln r + D \quad r > a$$

for a source–sink distribution $f(r) = \frac{1}{4}k(kr^2 - 1)\exp(-kr^2)(r < a)$, $f(r) = 0$ $(r > a)$, k is a constant. Use the boundary condition at the surface of the cylinder and the condition $\psi(r) \to 0$ as $r \to \infty$ to determine some of the arbitrary coefficients A, B, C, D. What will be the charge distribution on the surface of the cylinder?

2 Determine the locus of constant potential in the vicinity of an infinitely long vortex filament. How do the radial and angular velocities vary for a particle moving in the fluid field of a vortex filament?

3 Calculate the distribution of potential between an earthed conducting cylinder of infinite length, radius a, whose axis is parallel to and distance d away from a conducting sheet held at potential V_0 with respect to the cylinder. Determine also the charge distribution induced on the surface of the cylinder.

4 Show that the capacitance per unit length of two parallel cylindrical conductors of radius a and separation d is

$$C = \pi\varepsilon/\ln\left\{\frac{d}{2a}\sqrt{\left[\left(\frac{d}{2a}\right)^2 - 1\right]}\right\}$$

5 An alternating current of frequency 100 Mhz is passed along a non-magnetic conductor of diameter 1 mm. The conductivity is 10^7 ohm cm^{-1}. Establish the differential equation describing the radial distribution of current $j(r)$, and obtain an approximate solution. Over what frequency range would you expect this solution to be reasonably accurate.

6 A concentric cylindrical condenser of radii 100 mm and 102 mm is placed with one end in a liquid of dielectric constant ε. A steady potential of 300 V is applied across the plates. Neglecting capillary effects calculate the height the liquid rises inside the condenser. If now a small hole is drilled below the meniscus at a point A discuss the subsequent motion of the liquid through the system.

7 An infinitely long cylinder of radius a at potential V_0 has its axis parallel and distance d $(a \ll d)$ from a large earthed conducting sheet which is forced to vibrate as $a(t) = a_0 \sin \omega t$ along the normal passing through the axis of the cylinder. Determine the electrostatic energy developed between the cylinder and the plate averaged over one cycle of oscillation. How and why does it differ from the non-vibrating value? What can be said about the associated magnetic fields?

8 A uniform current i flows along a conducting cylinder of radius a whose axis is located at $x = b$ $(b > a)$ and is parallel to the yz plane. Assuming that the thermal conductivity of the cylinder and the surrounding medium is k, determine the equilibrium temperature distribution inside and outside the cylinder, given that the yz plane is kept at zero temperature.

9 Show that the velocity potential for an incompressible irrotational fluid flow about an infinitely long cylinder of radius a whose axis is perpendicular to the flow is

$$\psi(r, \theta) = v_0 r \cos\theta\left(1 + \frac{a^2}{r^2}\right)$$

and the corresponding velocity distribution is

$$v(r, \theta) = v_0\left\{\hat{r}\left(1 - \frac{a^2}{r^2}\right)\cos\theta - \hat{\theta}\left(1 + \frac{a^2}{r^2}\right)\sin\theta\right\}$$

where v_0 is the velocity of the fluid in the absence of the cylinder. Determine the kinetic energy of the fluid per unit length with and without the cylinder, and account for the difference physically.

10 The surface of an infinitely long cylinder of thermal conductivity k and radius a is maintained at a temperature T_0. A conductor of negligible cross-sectional area carries a current $a_0 e^{i\omega t}$ along the axis of the cylinder. Determine the equilibrium temperature distribution within the cylinder and comment on the distribution in the limits as ω becomes very large and very small. How would the solution be modified if the wire were placed eccentrically?

11 A pressure pulse $\delta(t)$ is applied uniformly over a circular membrane of radius a, density ρ. Neglecting the effect of the air load, show that the subsequent motion of the membrane is given as

$$\psi(r, t) = \frac{2a}{\rho c} \sum_m \frac{J_0(\pi k_m r/a)}{(\pi k_m)^2 J_1(\pi k_m a)} \sin(\pi k_m ct/a)$$

where c is the velocity of propagation of the wave and k_m is the mth root of $J_0(\pi k_m) = 0$. (Use the relationship $x J_0 = d(xJ_1)/dx$.)

12 A flat conducting strip of width a at potential V_0 is surrounded by a confocal elliptical insulating layer whose outer surface is at zero potential. If the resistivity of the material is p calculate the leakage current per unit length at equilibrium.

13 A coaxial transmission line consists of a thin-walled cylindrical metal tube of radius a and a cylindrical metal rod of radius b. The space between them is filled with an insulator of resistivity p. If the conductors are maintained at potentials V_a and V_b respectively show that the current flowing through the insulator is

$$i = \frac{2\pi}{p} \frac{V_b - V_a}{\log_e (a/b)}$$

per unit length.

A length l of the line is arranged so that one end of the rod and the whole outer surface of the tube are earthed. If a potential V is maintained across the conductors at the other end of the line show that the total current flowing through the insulator is

$$i = \frac{2\pi}{p_2 \log_e (a/b)} \frac{V}{\lambda} \left\{ \frac{\cosh(\lambda l) - 1}{\sinh(\lambda l)} \right\}$$

where

$$\lambda^2 = \frac{2p_1}{p_2} \frac{1}{b^2 \log_e (a/b)}$$

p_1 is the resistivity of the conductors, p_2 is the resistivity of the insulator and $p_2 \gg p_1$.

14 Show that the Fraunhofer diffraction pattern of a circular aperture of diameter a illuminated by a point source is given by

$$\frac{I(\theta)}{I(0)} = 4 \left\{ J_1 \left(\frac{\pi a \sin \theta}{\lambda} \right) \Big/ \left(\frac{\pi a \sin \theta}{\lambda} \right) \right\}^2$$

What would be the effect on this if the source were made larger and larger?

Discuss the design of a pin-hole camera to be used for taking X-ray pictures of the Sun (angular diameter 0·5°) at a wavelength of 10 Å. [The Bessel function $J_1(X)$ is given by

$$J_1(X) = \frac{X}{\pi} \int_0^\pi \cos(X \cos \alpha) \sin^2 \alpha \, d\alpha$$

CHAPTER 4

Orthogonal Curvilinear Coordinate Systems: Spherical Solutions

4.1   Introduction

We now come to consider the field distributions $\psi(r, \theta, \phi)$ in systems of spherical symmetry, and we shall generally expect to have to solve three single-variable differential equations for the radial $(R(r))$, polar $(\Phi(\phi))$ and azimuthal $(\Theta(\theta))$ components of the distribution, assuming the variables are separable. We shall certainly encounter a variety of radial solutions, depending upon the nature of the field equation, in particular the precise form of the radial potential function. The separated differential equations for $\Theta(\theta)$ and $\Phi(\phi)$ will, however, be largely independent of the details of the radial equation and the angular solutions therefore adopt a recurrent functional form. The angular distributions are termed the surface spherical harmonics $Y_l^m(\theta, \phi)$, where l and m are integers arising during the separation of the variables: to this extent the radial solution governs which values of l, and therefore which of the surface spherical harmonics, may develop. The integer values of m arise from the angular boundary condition that the distribution should be everywhere single-valued and continuous. Again, the condition on m that $l \geqslant m$ restricts the form of surface harmonics which can develop.

It is useful perhaps to regard the surface harmonics in much the same way as we regard the Fourier component description of a periodic function. Here, however, the distribution is periodic in θ and ϕ. We may, therefore, generally expand any angular distribution in terms of surface harmonic functions: their orthogonality properties will generally enable us to determine the harmonic coefficients.

A volume element in the system of spherical coordinates is shown in Figure 4.1.1. Laplace's equation in this system is

$$\frac{1}{r^2}\frac{\partial}{\partial r}\left(r^2\frac{\partial \psi}{\partial r}\right) + \frac{1}{r^2 \sin\theta}\frac{\partial}{\partial \theta}\left(\sin\frac{\partial \psi}{\partial \theta}\right) + \frac{1}{r^2 \sin^2\theta}\frac{\partial^2 \psi}{\partial \phi^2} = 0 \qquad (4.1.1)$$

If we assume a separation of variables of the form

$$\psi(r, \theta, \phi) = R(r)\Theta(\theta)\Phi(\phi) \qquad (4.1.2)$$

and substitute equation (4.1.2) in equation (4.1.1) and divide throughout by $R(r)\Theta(\theta)\Phi(\phi)$ we obtain

$$\frac{1}{r^2 R}\frac{d}{dr}\left(r^2\frac{dR}{dr}\right) + \frac{1}{r^2 \Theta \sin\theta}\frac{d}{d\theta}\left(\sin\theta\frac{d\Theta}{d\theta}\right) + \frac{1}{r^2 \sin^2\theta}\frac{d^2\Phi}{d\phi^2} = 0$$

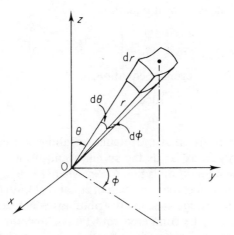

Figure 4.1.1 The volume element in spherical polar coordinates

Multiplication throughout by $r^2 \sin^2 \theta$ gives

$$\frac{\sin^2 \theta}{R} \frac{d}{dr}\left(r^2 \frac{dR}{dr}\right) + \frac{\sin \theta}{\Theta} \frac{d}{d\theta}\left(\sin \theta \frac{d\Theta}{d\theta}\right) + \frac{1}{\Phi} \frac{d^2\Phi}{d\phi^2} = 0 \qquad (4.1.3)$$

Setting

$$\frac{1}{\Phi} \frac{d^2\Phi}{d\phi^2} = -m^2 \qquad (4.1.4)$$

we have our first separated differential equation. Equation (4.1.3) then becomes, dividing throughout by $\sin^2 \theta$,

$$\frac{1}{R} \frac{d}{dr}\left(r^2 \frac{dR}{dr}\right) + \frac{1}{\Theta \sin \theta} \frac{d}{d\theta}\left(\sin \theta \frac{d\Theta}{d\theta}\right) - \frac{m^2}{\sin^2 \theta} = 0 \qquad (4.1.5)$$

and so we may effect a final separation

$$\frac{1}{R} \frac{d}{dr}\left(r^2 \frac{dR}{dr}\right) = l(l+1) \qquad (4.1.6)$$

and

$$\frac{1}{\Theta \sin \theta} \frac{d}{d\theta}\left(\sin \theta \frac{d\Theta}{d\theta}\right) + l(l+1) - \frac{m^2}{\sin^2 \theta} = 0 \qquad (4.1.7)$$

where the second separation constant has been written in the form $l(l+1)$. The three separated equations are, therefore,

$$\frac{d^2R}{dr^2} + \frac{2}{r} \frac{dR}{dr} - \frac{l(l+1)}{r^2} R = 0 \qquad (4.1.8)$$

which is termed Cauchy's linear equation,

$$\frac{d^2\Theta}{d\theta^2} + \frac{\cos\theta}{\sin\theta}\frac{d\Theta}{d\theta} + \left(l(l+1) - \frac{m^2}{\sin^2\theta}\right)\Theta = 0 \qquad (4.1.9)$$

which is the associated Legendre equation, and

$$\frac{d^2\Phi}{d\phi^2} + m^2\Phi = 0 \qquad (4.1.10)$$

We now see quite clearly that the angular solutions are dependent upon the radial solution only in as far as the specification of the separation constant $l(l+1)$. An ancillary condition is that $l \geqslant m$, where $l = 0, 1, 2, 3, \ldots$, but this is apparent only from the detailed solution of equation (4.1.9) (Appendix VI). The reader might like to try constructing distributions which violate the condition $l \geqslant m$. The results may be compared with those shown in Figure 4.3.1 which satisfy the condition.

The series solution of Cauchy's equation is straightforward. If we assume a solution of the form

$$R(r) = \sum_{n=-\infty}^{\infty} A_n r^n \qquad (4.1.11)$$

then equation (4.1.8) leads directly to the relation

$$\sum_{n=-\infty}^{\infty} [n(n-1) + 2n - l(l+1)]A_n r^{n-2} = 0$$

and this is non-trivially satisfied if

$$n(n-1) + 2n = l(l+1)$$

i.e. if

$$n = l, -(l+1) \qquad (4.1.12)$$

We therefore have the final general radial solution to the Laplace (Cauchy) equation:

$$R(r) = \sum_{l=0}^{\infty} \{B_l r^l + C_l r^{-(l+1)}\} \qquad (4.1.13)$$

The solution to equation (4.1.10) is

$$\Phi(m\phi) = A_m \sin m\phi + B_m \cos m\phi \qquad (4.1.14)$$

as in the cylindrical case. The condition that $\Phi(\phi)$ be everywhere continuous

$$\Phi = \Phi(\phi + 2\pi)$$

requires that m is an integer. The solutions to the $\Theta(\theta)$ equation (4.1.9) are the *associated Legendre functions* $P_l^m(\cos\theta)$, and their product with the $\Phi(m\phi)$ solutions yields the surface harmonics:

$$Y_l^m(\theta, \phi) = P_l^m(\cos\theta)\Phi(m\phi) \qquad (4.1.15)$$

If $m = 0$ so that there is no ϕ-dependence, then the Legendre functions reduce to their *unassociated* or *zonal* form, $P_l^0(\cos \theta)$. The unassociated functions are polynomials in $\cos \theta$ of degree l. Table 4.1.1 lists the first few unassociated Legendre functions for various l subject, of course, to the condition $m = 0$.

Table 4.1.1. Unassociated Legendre functions $P_l^0(\cos \theta)$

l	Mode	Polar coordinates	Cartesian coordinates
0	s	1	1
1	p_z	$\cos \theta$	z/r
2	d_z^2	$(3 \cos^2 \theta - 1)$	$(3z^2 - r^2)/r^2$
3	f_z^3	$(5 \cos^3 \theta - 3 \cos \theta)$	$(5z^3 - 3r^2z)/r^3$

These unassociated Legendre functions and their associated counterparts represent the general angular solution to problems of spherical symmetry. In particular the surface spherical harmonics describe the angular distribution of probability amplitude in the hydrogen atom and as such the first few modes are designated by their spectroscopic notation: s, p, d, f, ... corresponding to $l = 0, 1, 2, 3, \ldots$. The subscript z^n refers to the axis of symmetry of the distribution and the index n to the highest order of z in the cartesian representation. The unassociated functions listed in Table 4.1.1 are shown in Figure 4.1.2. Since there is no ϕ-dependence ($m = 0$) the distributions are circularly or rotationally symmetric about the z-axis. The s mode represents the quiescent spherically symmetric distribution, whilst the higher modes represent progressively more complex distributions.

We may now consider a specific problem in electrostatics. Suppose we wish to determine the electric field distribution in the region of an isolated hollow conducting sphere of radius a at potential ψ_0, subject to a uniform externally applied electric field E_0, directed along the positive z-axis (Figure 4.1.3). We might anticipate that there will be two solutions ψ_I and ψ_{II} corresponding to solutions inside and outside the sphere, respectively. Moreover the configuration of the problem suggests that there will be rotational symmetry about the z-axis. We may therefore assume the following forms of solution to the Laplace equation in the two regions:

$$\psi_I = \sum_{l=0}^{\infty} \{A_{Il}r^l + B_{Il}r^{-(l+1)}\}P_l^0(\cos \theta) \quad (r < a) \tag{4.1.16}$$

$$\psi_{II} = \sum_{l=0}^{\infty} \{A_{IIl}r^l + B_{IIl}r^{-(l+1)}\}P_l^0(\cos \theta) \quad (r > a) \tag{4.1.17}$$

where A_{II}, A_{III}, B_{II}, B_{III} are arbitrary coefficients to be determined from the boundary conditions. Since at large r ($\gg a$) the field will be effectively uniform it is clear that the potential must be of the form $-E_0r \cos \theta$—that is, we may assume that the angular distribution is described by the second unassociated

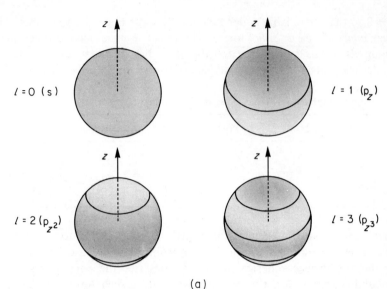

(a)

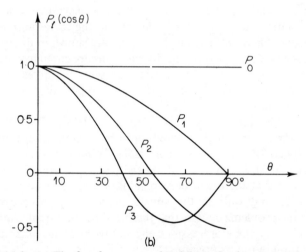

(b)

Figure 4.1.2 (a) The first four unassociated Legendre functions $P_l(\cos \theta)$ ($l = 0, 1, 2, 3$) tabulated in Table 4.1.1. All these distributions are symmetrical about the z-axis. (b) Graphical representation of the first four unassociated Legendre functions

surface harmonic $P_l^0(\cos \theta)(l = 1)$. And since we may reasonably assume a similar dependence in region I if we are to match the solutions at the boundary, then we have

$$\psi_{\text{I}} = (A_{\text{I}}r + B_{\text{I}}r^{-2})\cos \theta + C_{\text{I}} \tag{4.1.18}$$

$$\psi_{\text{II}} = (A_{\text{II}}r + B_{\text{II}}r^{-2})\cos \theta + C_{\text{II}} \tag{4.1.19}$$

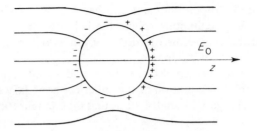

Figure 4.1.3 Distribution of induced charge on the surface of an earthed conducting sphere. The flux lines are incident normally on the surface of the sphere

where the function $P_l^0(\cos \theta) = \cos \theta$ (see Table 1.4.1) and C_{I} and C_{II} are constants of integration.

If the solution is to remain finite at the origin, then clearly $B_{\mathrm{I}} = 0$, whilst the boundary condition at $r = a$ yields

$$\psi_0 = A_{\mathrm{I}}a \cos \theta + C_{\mathrm{I}}$$

i.e.

$$A_{\mathrm{I}} = 0; \quad C_{\mathrm{I}} = \psi_0 \tag{4.1.20}$$

since ψ_0 is not a function of θ. Thus, within a conducting sphere there is a constant potential ψ_0.

In region II, far from the sphere the field becomes uniform and of magnitude $E_0 \cos \theta$. It therefore follows that

$$A_{\mathrm{II}} = -E_0$$

i.e.

$$\psi_{\mathrm{II}} = (-E_0 r + B_{\mathrm{II}} r^{-2}) \cos \theta + C_{\mathrm{II}} \tag{4.1.21}$$

and at the boundary,

$$\psi_0 = (-E_0 a + B_{\mathrm{II}} a^{-2}) \cos \theta + C_{\mathrm{II}}$$

so that

$$B_{\mathrm{II}} = E_0 a^3; \quad C_{\mathrm{II}} = \psi_0 \tag{4.1.22}$$

giving the final distributions of potential

$$\left.\begin{aligned} \psi_{\mathrm{I}} &= \psi_0 \\ \psi_{\mathrm{II}} &= E_0 \left[\frac{a^3}{r^2} - r\right] \cos \theta + \psi_0 \end{aligned}\right\} \tag{4.1.23}$$

The field inside the conducting sphere is evidently zero, whilst outside it is represented by the superposition of the uniform field plus that of a dipole: the sphere has acquired a dipole moment and has been polarized by the electric field. In the case of a conducting sphere the field distribution is as shown in Figure 4.1.3, and the charge distribution is given as $\varepsilon(\partial\psi/\partial r)_a = -3\varepsilon E_0 \cos\theta$, evaluated at the surface of the sphere.

A slightly more complicated electrostatic boundary value problem concerns the solution of Laplace's equation at the spherical boundary of two dielectric regions, Figure 4.1.4. Again we shall assume that the configuration is rotationally

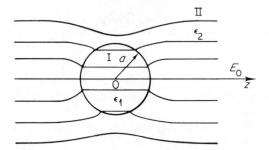

Figure 4.1.4 The internal and external flux distributions in the vicinity of a uniform dielectric sphere surrounded by a uniform dielectric

symmetric about the z-axis, and the angular dependence of the potential and field distributions will be described in terms of the unassociated surface spherical harmonics. The two regions are of dielectric constant ε_1 and ε_2, and a uniform field E_0 is applied along the positive z-axis. The general solution in either region has been shown to be (equation (4.1.16)) of the form

$$\psi(r, \theta) = \sum_{l=0}^{\infty} \{A_l r^l + B_l r^{-(l+1)}\} P_l^0(\cos\theta) \tag{4.1.24}$$

At large distances compared to the radius a of the sphere we require the field to regain its uniform value $E_0 \cos\theta$ which implies that the functional form of the potential distribution should be

$$\psi(r, \theta) = (Ar + Br^{-2})\cos\theta \quad (l = 1) \tag{4.1.25}$$

Adopting this form of solution in regions I and II we have from the boundary condition at infinity,

$$\psi_{\text{II}}(r, \theta) = (-E_0 r + B_{\text{II}} r^{-2})\cos\theta \tag{4.1.26}$$

The requirement that the potential should not diverge at the origin enables us to set $B_{\text{I}} = 0$, so that

$$\psi_{\text{I}}(r, \theta) = A_{\text{I}} r \cos\theta \tag{4.1.27}$$

and the two constants $A_{\text{I}}, B_{\text{II}}$ remain to be determined from the boundary conditions.

The tangential component of the field E_θ must be continuous across the boundary of the sphere, and so we have

$$\left(-\frac{1}{r}\frac{\partial\psi_{\mathrm{I}}}{\partial\theta}\right)_a = \left(-\frac{1}{r}\frac{\partial\psi_{\mathrm{II}}}{\partial\theta}\right)_a \qquad (4.1.28)$$

where these terms represent the angular components of the gradient operator at the dielectric boundary. Equation (4.1.28) yields

$$A_1 = -E_0 + \frac{B_{\mathrm{II}}}{a^3} \qquad (4.1.29)$$

We may also use the condition that the normal electric displacement vector **D** must be continuous across the boundary, that is,

$$\left(-\varepsilon_1\frac{\partial\psi_{\mathrm{I}}}{\partial r}\right)_a = \left(-\varepsilon_2\frac{\partial\psi_{\mathrm{II}}}{\partial r}\right)_a \qquad (4.1.30)$$

and this yields

$$\varepsilon_1 A_1 = -\varepsilon_2 E_0 - \frac{2\varepsilon_2 B_{\mathrm{II}}}{a^3} \qquad (4.1.31)$$

Equations (4.1.29) and (4.1.31) may be solved for A_1 and B_{II};

$$A_1 = -\frac{3\varepsilon_2 E_0}{\varepsilon_1 + 2\varepsilon_2} \qquad (4.1.32)$$

$$B_{\mathrm{II}} = (E_0 - A_1)a^3 \qquad (4.1.33)$$

$$= \left(\frac{\varepsilon_1 - \varepsilon_2}{\varepsilon_1 + 2\varepsilon_2}\right)a^3 E_0$$

We finally obtain the potential distributions

$$\left. \begin{aligned} \psi_{\mathrm{I}}(r,\theta) &= \frac{-3\varepsilon_2 E_0 r\cos\theta}{\varepsilon_1 + 2\varepsilon_2} \\[2mm] \psi_{\mathrm{II}}(r,\theta) &= E_0\left[\left(\frac{\varepsilon_1 - \varepsilon_2}{\varepsilon_1 + 2\varepsilon_2}\right)\frac{a^3}{r^2} - r\right]\cos\theta \end{aligned} \right\} \qquad (4.1.34)$$

In the limiting case of $\varepsilon_1 = \varepsilon_2$ we regain the perfectly uniform field $E_0\cos\theta$, whilst if $\varepsilon_1 \gg \varepsilon_2$ the distribution of potential approaches that of the grounded conducting sphere (equation (4.1.23) with $\psi_0 = 0$).

The equation for the distribution $\psi_{\mathrm{I}}(r,\theta)$ may be rewritten in a particularly instructive way, as follows:

$$\psi_{\mathrm{I}}(r,\theta) = -E_0 r\cos\theta + \frac{(\varepsilon_1 - \varepsilon_2)}{2\varepsilon_2 + \varepsilon_1}E_0 r\cos\theta \qquad (4.1.35)$$

In other words the field within the dielectric sphere may be regarded as a super-position of the external polarizing field plus a *depolarizing* term. The positive or

130

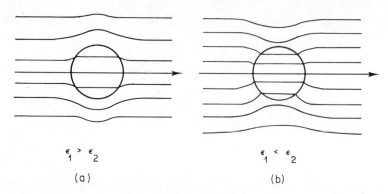

$\epsilon_1 > \epsilon_2$
(a)

$\epsilon_1 < \epsilon_2$
(b)

Figure 4.1.5 Schematic internal and external distributions of flux in the vicinity of a uniform sphere of dielectric constant ε_1, surrounded by a uniform medium of dielectric constant ε_2 (a) $\varepsilon_1 > \varepsilon_2$ (b) $\varepsilon_1 < \varepsilon_2$. In the former case lines are 'expelled' whilst in the latter they are 'attracted' into the sphere. A precisely analogous situation exists for magnetic media of permeabilities μ_1 and μ_2

negative direction of this depolarizing field depends on the relative magnitudes of ε_1 and ε_2. The two situations are shown in Figure 4.1.5.

Suppose we now compute the field about a flat circular disc of radius a with its normal axis in the z-direction (Figure 4.1.6). The disc is uniformly charged to Q so that the charge density is $Q/\pi a^2$. The potential at a point P on the z-axis distance r from the origin will of course be given as

$$\psi(\theta = 0) = \frac{Q}{\pi a^2} \cdot 2\pi \int_0^a \frac{x \, dx}{\sqrt{(r^2 + x^2)}} = \frac{2Q}{a^2} \{\sqrt{(a^2 + r^2)} - r\}$$

$$= \frac{2Q}{a} \left\{ 1 - \left(\frac{r}{a}\right) + \frac{1}{2}\left(\frac{r}{a}\right)^2 - \frac{1}{8}\left(\frac{r}{a}\right)^4 - \frac{1}{16}\left(\frac{r}{a}\right)^6 - \cdots \right\} \quad (r < a)$$

$$= \frac{2Q}{a} \left\{ \frac{1}{2}\left(\frac{a}{r}\right) - \frac{1}{8}\left(\frac{a}{r}\right)^3 + \frac{1}{16}\left(\frac{a}{r}\right)^5 - \cdots \right\} \quad (r > a)$$

$$(4.1.36)$$

where the expansions follow directly from the binomial theorem. It is evident from the symmetry of the problem that the distribution is cylindrically symmetric about the z-axis and the solution at the general point $P'(r, \theta, \phi)$ may be expressed in terms of the zonal harmonics

$$\sum_{l=0}^{\infty} P_l^0(\cos \theta) r^l \quad \text{or} \quad \sum_{l=0}^{\infty} P_l^0(\cos \theta) r^{-(l+1)}$$

(cf. equations (4.1.16) and (4.1.17)). These solutions are chosen so as to remain finite at the origin and infinity, respectively. It therefore follows that by fitting the appropriate zonal series to the two radial expansions (4.1.36) we obtain the

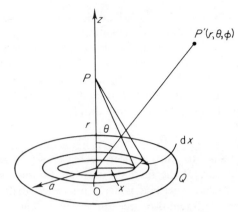

Figure 4.1.6 Geometry for the calculation of the distribution of potential about a flat circular disc of uniform total charge Q

complete description

$$\psi(r, \theta, \phi) = \frac{2Q}{a} \left\{ 1 - \left(\frac{r}{a}\right) P_1(\cos \theta) + \frac{1}{2}\left(\frac{r}{a}\right)^2 P_2(\cos \theta) - \frac{1}{8}\left(\frac{r}{a}\right)^4 P_4(\cos \theta) + \cdots \right\}$$

$$(r < a)$$

$$= \frac{2Q}{a} \left\{ \frac{1}{2}\left(\frac{a}{r}\right) - \frac{1}{8}\left(\frac{a}{r}\right)^3 P_2(\cos \theta) + \frac{1}{16}\left(\frac{a}{r}\right)^5 P_4(\cos \theta) - \cdots \right\} \quad (r > a)$$

$$(4.1.37)$$

provided the $P_l^0(\cos \theta)$ are normalized such that $P_l^0(0) = 1$. The solution is independent of ϕ, as indeed it should be, and at large distances the potential reduces to Q/r, as it should.

This is, in fact, an application of the more general *axis theorem*. If a potential function for a system of axial symmetry may be expressed in the form

$$\psi(z) = \sum_{l=0}^{\infty} \{A_l z^l + B_l z^{-(l+1)}\}$$

at a point on the axis, then the potential at the point (r, θ, ϕ) is

$$\psi(r, \theta, \phi) = \sum_{l=0}^{\infty} \{A_l r^l + B_l r^{-(l+1)}\} P_l^0(\cos \theta)$$

Cole describes a geomagnetic application of the general solution

$$\psi(r, \theta, \phi) = \sum_l \sum_m \{A_{lm} r^l + B_{lm} r^{-(l+1)}\} Y_l^m(\theta, \phi) \qquad (4.1.38)$$

If we suppose that the earth's magnetic field is derivable from the above potential function, and we further suppose that the earth is a sphere of radius a, then a measurement of the magnetic field at the earth's surface will enable us to determine the coefficients A_{lm}, B_{lm}. Clearly, if A_{lm} is zero the field is of internal origin and only the external region of the solution will appear. The possibility thus is that the existence of electric currents between the earth's surface and the atmosphere will yield fields which cannot be described by a potential of the form (4.1.35). In fact it is found that the A_{lm} are very nearly zero, indicating an internal origin to the earth's magnetic field, and the dominating coefficient is $B_{1,0}$ corresponding to the potential due to a dipole:

$$\psi \sim B_{1,0} r^{-2}$$

The coefficient $B_{1,0}$ seems to show a slow decrease, $\sim 6\%$ per century. The secular or non-dipole part of the field shows a continuous change over the surface of the earth together with a general westward drift which appears uncorrelated with any continental or oceanic features. The internal origin of the field is believed to be associated with magnetohydrodynamic electric currents in the liquid core of the earth.

4.2   Uniformly Moving Sphere in an Ideal Fluid

By an ideal fluid we mean a continuous incompressible system which offers no resistance to motion, the only force exerted on a body being a hydrostatic pressure. Thus we are considering the fluid to be non-viscous, and in considering it to be incompressible so that the density remains constant, we are stating the continuity condition

$$\text{div } \mathbf{v} = 0 \tag{4.2.1}$$

where $\mathbf{v}$ is the velocity vector of the fluid flux. The fluid flow is irrotational (see Chapter 1), and we may therefore assume a relation

$$\mathbf{v} = -\text{grad } \psi \tag{4.2.2}$$

where ψ is the *velocity potential*. From equations (4.2.1) and (4.2.2) Laplace's equation for the velocity potential follows immediately

$$\text{div grad } \psi = \nabla^2 \psi = 0 \tag{4.2.3}$$

We shall be concerned with determining the *streamline* of the fluid flow, and this is defined as the locus which has at every point the same direction as the velocity vector $\mathbf{v}$. It then follows that the equipotential surfaces are everywhere perpendicular to the streamlines.

Suppose a sphere of radius a is moving uniformly along the z-direction with a velocity V in an incompressible fluid at rest. We wish to determine the distribution of fluid velocity $v(r, \theta)$ at some arbitrary point P in the fluid in the vicinity of the sphere, Figure 4.2.1. Quite clearly the distribution will be rotationally symmetric about the z-axis, and the Laplace equation may consequently be written

$$\frac{\partial}{\partial r}\left(r^2 \frac{\partial \psi}{\partial r}\right) + \frac{1}{\sin \theta}\frac{\partial}{\partial \theta}\left(\sin \theta \frac{\partial \psi}{\partial \theta}\right) = 0 \tag{4.2.4}$$

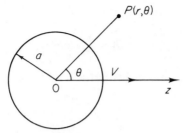

Figure 4.2.1 Geometry for the calculation of the distribution of velocity potential of an irrotational, incompressible fluid about a fixed sphere, radius a

Now, the *radial* component of the fluid velocity v_r at the surface of the sphere must be

$$(v_r)_a = -\left(\frac{\partial \psi}{\partial r}\right)_a = V \cos \theta \qquad (4.2.5)$$

if the fluid is not to penetrate the sphere. Our other boundary condition is simply that at infinity the radial velocity component must be zero, i.e. the fluid must be at rest:

$$(v_r)_\infty = -\left(\frac{\partial \psi}{\partial r}\right)_\infty = 0 \qquad (4.2.6)$$

Equations (4.2.5) and (4.2.6) constitute the Neumann boundary conditions to the problem. Assuming a separable solution of the form $\psi(r, \theta) = R(r)\Theta(\theta)$, equation (4.2.4) separates into the two equations (4.1.8), (4.1.9). As we have seen, the general solution to equation (4.2.4) may be written (cf. equation (4.1.17))

$$\psi(r, \theta) = \sum_{l=0}^{\infty} \{A_l r^l + B_l r^{-(l+1)}\} P_l^0(\cos \theta) \qquad (4.2.7)$$

However, if we are to accommodate the boundary condition (4.2.5), it is evident that the particular form of solution for the velocity potential is

$$\psi(r, \theta) = (Ar + Br^{-2}) \cos \theta \quad (l = 1) \qquad (4.2.8)$$

where A and B are arbitrary constants to be determined from the boundary conditions. We immediately have

$$\frac{\partial \psi}{\partial r} = \left(A - \frac{2B}{r^3}\right) \cos \theta \qquad (4.2.9)$$

and from equation (4.2.6) it follows that $A = 0$. The other boundary condition (4.2.5) yields

$$-\left(\frac{\partial \psi}{\partial r}\right)_a = \frac{2B}{a^3} \cos \theta = V \cos \theta \qquad (4.2.10)$$

so that $B = \frac{1}{3}Va^2$, and the velocity potential is

$$\psi(r, \theta) = \frac{Va^3}{2r^2} \cos \theta \tag{4.2.11}$$

The radial and angular components of the fluid velocity are, respectively,

$$\left. \begin{array}{l} v_r = -\dfrac{\partial \psi}{\partial r} = V \left(\dfrac{a}{r}\right)^3 \cos \theta \\[3mm] v_\theta = -\dfrac{1}{r} \dfrac{\partial \psi}{\partial \theta} = \dfrac{V}{2} \left(\dfrac{a}{r}\right)^3 \sin \theta \end{array} \right\} \tag{4.2.12}$$

and the streamlines may be obtained as orthogonal surfaces to the equipotentials, equation (4.2.11) (Figure 4.2.2). This follows as a consequence of the definition (4.2.2).

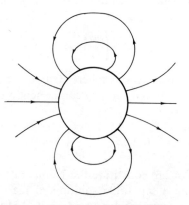

Figure 4.2.2 The streamlines associated with the uniform flow about a sphere of radius a held fixed with respect to a fluid of initial velocity **v**. The streamlines are orthogonal surfaces to the velocity equipotentials (4.2.11)

We may easily calculate the total kinetic energy of the system as follows:

$$\text{k.e. (fluid)} = \frac{2\pi\rho}{2} \int_a^\infty \int_0^\pi v(r, \theta)^2 \, r^2 \sin \theta \, d\theta \, dr \tag{4.2.13}$$

where ρ is the fluid density and $r^2 \sin \theta \, d\theta \, dr$ is a volume element. $v^2 = (v_r^2 + v_\theta^2)$. From equation (4.2.12) the integral may be rewritten:

$$\pi\rho V^2 a^6 \int_a^\infty \left[\int_0^\pi (\cos^2 \theta + \tfrac{1}{4} \sin^2 \theta) \sin \theta \, d\theta \right] \frac{1}{r^4} \, dr$$

$$= \tfrac{1}{4}(\tfrac{4}{3}\pi a^3 \rho) V^2 = \tfrac{1}{4} m_f V^2 \tag{4.2.14}$$

m_f is the mass of the fluid displaced by the sphere. The total kinetic energy (sphere + fluid) is therefore

$$\text{k.e.} = \tfrac{1}{2}mV^2 + m_f V^2$$
$$= \tfrac{1}{2}m^*V^2 \qquad (4.2.15)$$

where $m^* = m + \tfrac{1}{2}m_f$ and m is the mass of the sphere.

Thus the sphere moves as if it has an *effective mass* m^*, greater than its real mass. This is understood in terms of the kinetic energy imparted to the fluid in moving the sphere.

4.3   Vibrational Modes of a Spherical Droplet

If we consider the equation of continuity for an incompressible fluid in conjunction with the equation of motion of a volume element in spherical coordinates we obtain the wave equation describing the propagation of surface waves over the droplet, provided the amplitude $\psi(\theta, \phi, t)$ is small in comparison with its radius. Thus,

$$\frac{1}{R^2 \sin \theta} \frac{\partial}{\partial \theta}\left(\sin \theta \frac{\partial \psi}{\partial \theta}\right) + \frac{1}{R^2 \sin^2 \theta} \frac{\partial^2 \psi}{\partial \phi^2} = \frac{1}{c^2} \frac{\partial^2 \psi}{\partial t^2} \qquad (4.3.1)$$

where R is the radius of the unexcited spherical droplet and c is the velocity of propagation of the wave across the liquid. There are two distinct cases according to whether the waves are short wavelength surface capillary waves or long wavelength oceanic tidal waves. In the former case the propagation process is largely governed by the surface tension and density ($c^2 = \gamma/\rho$), whilst for surface tidal waves on a planet of radius R the velocity of propagation is governed by the depth of the ocean and the acceleration due to gravity ($c^2 = gh$). In the case of the surface capillary waves we have assumed that they are non-dispersive which is not strictly true. However, for the purposes of the present discussion we shall assume c is a constant independent of the wavelength.

We notice first of all that there are no boundary conditions as such, and the only condition on the field $\psi(\theta, \phi, t)$ is that it should be everywhere a single valued and continuous function. As we have seen before, this is sufficient to restrict which normal modes of vibration the system is able to sustain, and arises directly from the requirement that the function must satisfy

$$\left.\begin{array}{l} \Theta(\theta) = \Theta(\theta + 2\pi) \\ \Phi(\phi) = \Phi(\phi + 2\pi) \end{array}\right\} \qquad (4.3.2)$$

Thus the frequency of the normal mode solutions will be a discrete and not a continuous distribution. If we assume a separable solution of the form

$$\psi(\theta, \phi, t) = \Theta(\theta)\Phi(\phi)T(t) \qquad (4.3.3)$$

Then we immediately have

$$T(t) = A \sin(2\pi\nu t + \delta) \qquad (4.3.4)$$

where δ is a phase constant, and v is the frequency of vibration. Equation (4.3.1) may be immediately rewritten

$$\frac{1}{R^2\Theta(\theta)\sin\theta}\frac{d}{d\theta}\left(\sin\theta\frac{d\Theta}{d\theta}\right) + \frac{1}{R^2\Phi(\phi)\sin^2\theta}\frac{d^2\Phi}{d\phi^2} + \frac{4\pi^2 v^2}{c^2} = 0 \quad (4.3.5)$$

where the separated expression of the field, (4.3.3), has been used. Multiplying equation (4.3.5) throughout by $R^2\sin^2\theta$ we obtain the two separated equations

$$\frac{1}{\sin\theta}\frac{d}{d\theta}\left(\sin\theta\frac{d\Theta}{d\theta}\right) + \left\{\frac{4\pi^2 v^2 R^2}{c^2} - \frac{m^2}{\sin^2\theta}\right\}\Theta = 0 \quad (4.3.6)$$

$$\frac{d^2\Phi}{d\phi^2} + m^2\Phi = 0 \quad (4.3.7)$$

where m^2 is a separation constant. Equation (4.3.7) has the general solution

$$\Phi(\phi) = B\sin m\phi + C\cos m\phi \quad (4.3.8)$$

where B and C are arbitrary constants. Only two solutions to equation (4.3.7) are linearly independent, and it is convenient for us to take these as $\sin m\phi$ and $\cos m\phi$. From the condition (4.3.2) it follows that m must be an integer, and may be zero.

Slight rearrangement of equation (4.3.6) shows it to be identical to equation (4.1.9) provided we make the identification

$$l(l+1) = \frac{4\pi^2 v^2 R^2}{c^2} \quad (4.3.9)$$

where for finite valued solutions l is an integer $l = 0, 1, 2, 3, \ldots$ and $l \geq |m|$. We therefore see that there is a *discrete* set of normal mode frequencies given by

$$v_l = \frac{c}{2\pi R}\sqrt{[l(l+1)]} \quad (4.3.10)$$

and these are independent of the parameter m. It therefore follows that each normal mode is $(2l+1)$-fold degenerate since $|m|$ may take on values 1 to l, each doubly degenerate (see equation (4.3.8)), plus the $m = 0$ mode. We also see that the harmonics are not simple multiples of the fundamental frequency, $v_1 = c\sqrt{2}/2\pi R$, corresponding to $l = 1$. The mode $l = 0$ has zero frequency and represents the neutral or unexcited spherically symmetric mode.

As we observed in the case of the angular solutions to the Laplace equation, the general solution is expressed in terms of a linear combination of surface spherical harmonics, $Y_l^m(\theta, \phi)$:

$$\sum_{l=0}^{\infty}\sum_{m=-l}^{l} Y_l^m(\theta, \phi) = \sum_{l=0}^{\infty}\sum_{m=-l}^{l} A_{lm}P_l^m(\cos\theta)\Phi(m\phi) \quad (4.3.11)$$

where $P_l^m(\cos\theta)$ are the *associated Legendre functions*, and $\Phi(m\phi)$ is the trigonometric solution (4.3.8). A_{lm} are arbitrary harmonic coefficients. The unassociated spherical harmonics having $m = 0$ are also termed the *zonal harmonics* (since these functions depend only on θ, the nodal lines divide the sphere into zones—see Figure 4.1.1). The two other classes of associated surface harmonic functions which we have not yet discussed are the *sectoral harmonics* ($l = m$, the nodal lines dividing the spherical surface into sectors) and the remainder which are termed the *tesseral harmonics* ($-l < m < l$). These functions are the eigenfunctions for the two-dimensional spherical surface, and are mutually orthogonal functions.

The first few associated Legendre functions are listed in Table 4.3.1, and are shown in Figure 4.3.1. There are also the associated Legendre functions of the second kind, Q_l^m, having an imaginary argument, but we shall not encounter these until we develop the bi-centric spheroidal coordinates (Section 4.14). They may be physically rejected here since they do not lead to finite solutions.

Table 4.3.1. Associated spherical harmonics $Y_l^m(\cos \theta)$

l	m	Mode	Polar coordinates	Cartesian coordinates
0	0	s	1	1
1	0	p_z	$\cos \theta$	z/r
1	1	$\begin{cases} p_y \\ p_x \end{cases}$	$\begin{cases} \sin \theta \sin \phi \\ \sin \theta \cos \phi \end{cases}$	$\begin{matrix} y/r \\ x/r \end{matrix}$
2	0	d_{z^2}	$3\cos^2 \theta - 1$	$(3z^2 - r^2)/r^2$
2	1	$\begin{cases} d_{yz} \\ d_{xz} \end{cases}$	$\begin{cases} \sin \theta \cos \theta \sin \phi \\ \sin \theta \cos \theta \cos \phi \end{cases}$	$\begin{matrix} yz/r^2 \\ xz/r^2 \end{matrix}$
2	2	$\begin{cases} d_{xy} \\ d_{x^2-y^2} \end{cases}$	$\begin{cases} \sin^2 \theta \sin 2\phi \\ \sin^2 \theta \cos 2\phi \end{cases}$	$\begin{matrix} xy/r^2 \\ (x^2 - y^2)/r^2 \end{matrix}$
3	0	f_{z^3}	$5\cos^3 \theta - 3\cos \theta$	$(5z^3 - 3r^2 z)/r^3$
3	1	$\begin{cases} f_{yz^2} \\ f_{xz^2} \end{cases}$	$\begin{cases} \sin \theta (5\cos^2 \theta - 1) \sin \phi \\ \sin \theta (5\cos^2 \theta - 1) \cos \phi \end{cases}$	$\begin{matrix} y(5z^2 - r^2)/r^3 \\ x(5z^2 - r^2)/r^3 \end{matrix}$
3	2	$\begin{cases} f_{xyz} \\ f_{z(x^2-y^2)} \end{cases}$	$\begin{cases} \sin^2 \theta \cos \theta \sin 2\phi \\ \sin^2 \theta \cos \theta \cos 2\phi \end{cases}$	$\begin{matrix} xyz/r^3 \\ z(x^2 - y^2)/r^3 \end{matrix}$
3	3	$\begin{cases} f_{y^3} \\ f_{x^3} \end{cases}$	$\begin{cases} \sin^3 \theta \sin 3\phi \\ \sin^3 \theta \cos 3\phi \end{cases}$	$\begin{matrix} y(y^2 - 3x^2)/r^3 \\ x(x^2 - 3y^2)/r^3 \end{matrix}$

As we have already observed, the frequency of a given normal mode ν_l is a degenerate function of the degree l of the Legendre function, equation (4.3.10). There is therefore the possibility of extensive hybridization amongst the normal modes, particularly those of the same degree. (We recall that hybridization or coupling between independent normal modes of vibration occurs most strongly amongst normal modes of similar frequency). Thus if we form hybrids from a

138

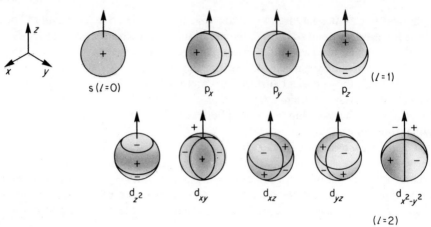

Figure 4.3.1 The first few associated Legendre distributions $P_l^m(\cos\theta)$ tabulated in Table 4.3.1

linear combination of p modes we obtain the rotated distributions shown in Figure 4.3.2. We may easily understand how the distribution arises by considering the cartesian subscripts. The $p_x + p_y$ hybrid, for example, has its maximum at 45° to the positive x- and y-axes in the xy plane. The $p_x + p_y + p_z$ hybrid has its maximum symmetrically placed between the x-, y- and z-axes. Clearly the general hybrid $ap_x + bp_y + cp_z$ may be used to represent any orientation of the distribution, depending upon the relative magnitudes of the hybridizing coefficients a, b, c.

Degenerate normal modes may also be combined with different phases in which case the nodal lines are no longer stationary but move about over the vibrating surface. This of course applies to any system executing a hybrid vibration, and is not specifically restricted to the case of surface spherical harmonics. If, for example we take the surface harmonic hybrid $(a(t)\,d_{x^2} + b(t)\,d_{y^2})$ where the hybridization coefficients are one quarter cycle out of phase, then we have

$$Y_3(\theta, \phi, t) = \cos\left(\frac{2\pi t}{\tau}\right) d_{x^2} + \sin\left(\frac{2\pi t}{\tau}\right) d_{y^2} \qquad (4.3.12)$$

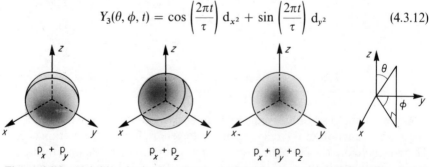

Figure 4.3.2 Hybrids obtained as linear combinations of the associated functions shown in Figure 4.3.1 and listed in Table 4.3.1. Note how the distribution is effectively rotated, but otherwise remains unchanged

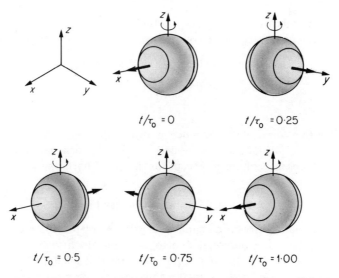

Figure 4.3.3 The hybrid $a(t)\,d_{x^2} + b(t)\,d_{y^2}$ where the coefficients $a(t)$, $b(t)$ vary sinusoidally $\pi/4$ out of phase. This is seen to effectively rotate the distribution about the z-axis. Such a hybrid of normal modes describes the effect of the moon's motion on the earth's tides, and accounts for the existence of two high tides each day, even though the moon's period is 24 hours

where $\tau_0 = 2\pi R/c\sqrt{6}$ is the period of oscillation. The movement of the nodal lines across the surface is shown in Figure 4.3.3, and is seen to result in the rotation of a zonal d mode about the z-axis with period τ_0. In the case of oceanic waves over the earth's surface, the phase retarded hybrid (4.3.12) quite closely describes the terrestrial tidal motions due to the moon, including the existence of two high tides per day at a given point on the earth's surface even though the moon rotates around the earth only once every 24 hours.

If we now consider the normal modes of vibration of a *sessile* droplet in a gravitational field, acting in the negative z-direction, then we find that the $(2l + 1)$-fold degeneracy of the lth normal mode is partially removed. We assume that the gravitational distortion of the droplet is small so that we can correctly correlate normal modes of the same degree in the distorted and undistorted systems. From Figure 4.3.4 it is quite clear that the three sessile p modes *cannot*

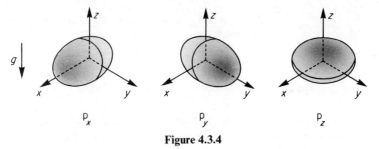

Figure 4.3.4

have the same vibrational frequency, although the p_x and p_y modes remain degenerate. If we examine the trigonometric expression of these functions

$$
\left.\begin{array}{l}
p_x = \sin\theta\cos\phi \\
p_y = \sin\theta\sin\phi \\
p_z = \cos\theta
\end{array}\right\}
\left\{\begin{array}{l}
l = 1, m = 1 \\
l = 1, m = 1 \\
l = 1, m = 0
\end{array}\right.
$$

we see that the vibrational frequency is now dependent upon *both* l and m (cf. equation (4.3.10)) and each l, m mode is now doubly degenerate (corresponding to $\sin m\phi$, $\cos m\phi$ solutions) except for the zonal harmonics, $m = 0$.

4.4 Electrostatic Multipolar Fields

A further simple electrostatic example of the use of the spherical harmonic functions concerns the potential due to an arbitrary distribution of charge density within a sphere radius a, at a point outside this sphere. We have already discussed the solution to the Laplace equation in spherical coordinates, but restricted the angular distribution to the unassociated (zonal) Legendre functions. It is quite apparent that the appropriate generalization of equation (4.1.7) to include the associated harmonics results in the degenerate distribution

$$
\psi(r, \theta, \phi) = \sum_{l=0}^{\infty} \sum_{m=-l}^{l} A_{lm} P_l^m(\cos\theta)\Phi(m\phi)r^{-(l+1)} \quad (r > a) \tag{4.4.1}
$$

this remaining finite at infinity (cf. equation (4.1.17)).

If we now expand equation (4.4.1) we obtain the potential at an arbitrary point $(r > a)$ as

$$
\psi(r, \theta, \phi) = A_{00}Y_0^0(\theta, \phi)\frac{1}{r} + (A_{10}Y_1^0(\theta, \phi) + 2A_{11}Y_1^1(\theta, \phi))\frac{1}{r^2}
$$
$$
\tag{4.4.2}
$$
$$
+ (A_{20}Y_2^0(\theta, \phi) + 2A_{21}Y_2^1(\theta, \phi) + 2A_{22}Y_2^2(\theta, \phi))\frac{1}{r^3} + \cdots
$$

or, directly in terms of the harmonic modes:

$$
= \frac{A_{00}}{r} + (A_{1x}p_x + A_{1y}p_y + A_{1z}p_z)\frac{1}{r^2}
$$
$$
+ (A_{2z^2}d_{z^2} + A_{2xz}d_{xz} + A_{2yz}d_{yz} + A_{2xy}d_{xy} + A_{2x^2-y^2}d_{x^2-y^2})\frac{1}{r^3}
$$
$$
+ (A_{3z^3}f_{z^3} + A_{3yz^2}f_{yz^2} + A_{3x^2}f_{xz^2} + A_{3xyz}f_{xyz} \tag{4.4.3}
$$
$$
+ A_{3z(x^2-y^2)}f_{z(x^2-y^2)} + A_{3y^3}f_{y^3} + A_{3x^3}f_{x^3})\frac{1}{r^4} + \cdots
$$

This *multipole* expansion of the potential distribution has a direct physical interpretation. Clearly the first term in the series represents the potential at the point r due to the net charge at the origin of the cluster, whilst the second term represents the correction to this distribution at the origin due to any net *dipole* moment the cluster might possess. The three cartesian orientations of the dipole and any hybrid of these charge distributions are accounted for by the coefficients A_{1x}, A_{1y}, A_{1z}. Similarly, the third term describes the *quadrupole* components of the potential distribution, the fourth the *octopole* moments, and so on. It is evident that the dipole, quadrupole, octopole, etc., terms are represented by the degenerate set of p, d, f, etc., spherical harmonics, and the first few of these are shown in Figure 4.4.1 (cf. Figure 4.3.1).

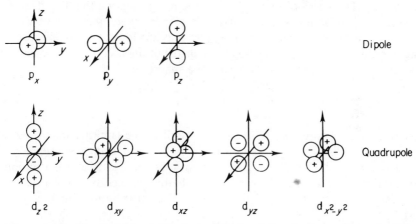

Figure 4.4.1 The first few multipoles and their relation to the surface spherical harmonics

The quadrupole term has five components and both it and subsequent multipolar components do not, therefore, behave like a vector. It should be pointed out that the spherical harmonic fields generated by a given term in the series cannot be exactly generated from combinations of other multipolar distributions. It follows that as far as the potential distribution outside the large cluster is concerned, *any charge distribution may be replaced by a suitable assembly of multipoles.* This should not surprise us. We are perfectly aware of the Fourier representation of a linear periodic function by a suitable linear combination of orthogonal trigonometric functions: the present expansion may be regarded as the angular analogue in θ–ϕ space. The functions are angularly periodic, they are orthogonal, and collectively they constitute a *complete set* of functions suitable for a harmonic expansion of the type we have just described.

These multipolar expansions form a useful basis for the discussion of the electrostatic potential in the vicinity of small molecular clusters, such as that of the water molecule. In considering a condensed assembly of such molecules, a systematic account of the electrostatic component of the interaction energies may be made by the progressive inclusion of higher and higher multipole terms.

The problem is, of course, one of determining the multipole tensors A_{lm} in equation (4.4.2).

We shall finally very briefly mention the form of the potential distribution *inside* a sphere of radius a, having the potential $\psi_a(\theta, \phi)$ specified on its surface. As we have seen, the appropriate generalization of equation (4.1.16) such that the solution remains finite at the origin is

$$\psi(r, \theta, \phi) = \sum_{l=0}^{\infty} \sum_{m=-l}^{l} A_{lm} Y_l^m(\theta, \phi) \left(\frac{r}{a}\right)^l \quad (r < a) \tag{4.4.4}$$

where the arbitrary coefficients A_{lm} are to be determined from the boundary conditions. The terms of high degree (large l) representing the 'fine details' of the surface distribution ψ_a are clearly only of importance when r is very nearly equal to a. Nearer the centre of the sphere only the lower harmonics affect the angular distribution and the potential distribution becomes more uniform. The corresponding distribution *outside* the sphere becomes

$$\psi(r, \theta, \phi) = \sum_{l=0}^{\infty} \sum_{m=-l}^{l} A_{lm} Y_l^m(\theta, \phi) \left(\frac{r}{a}\right)^{-(l+1)} \quad (r > a) \tag{4.4.5}$$

ensuring that the potential goes to zero at infinity. The same argument suggests that at large distances the spherical harmonic field becomes progressively indifferent to the high-order angular fine structure of the surface distribution of potential, $\psi_a(\theta, \phi)$. Ultimately $(r \gg a)$ the potential reduces to that of a Coulomb field varying as $1/r$, when the details of the surface distribution can no longer be resolved.

4.5 Normal Interior Modes of Vibration of a Rotating Gaseous Star

So far we have considered no radial solutions to the wave equation in spherical coordinates, and here we attempt to determine the distribution of pressure $\psi(r, \theta, \phi)$ in the body of a spherical gaseous star rotating about the z-axis which is subject to internal gravitational and centrifugal forces, but which nevertheless has a uniform distribution of temperature and density. Any discussion of the departure from sphericity which will inevitably occur in the rotating system will have to be deferred until we have considered the spheroidal coordinate system. The actual pressure variation within a rotating gaseous star will depend principally upon the gravitational and centrifugal terms, the former predominating for a stable stellar system. Various models have been proposed, and in Figure 4.5.1 we show the schematic pressure distribution within the photosphere of our own sun.

A consideration of the equation of continuity and the equation of motion of a volume element of the star yields the wave equation describing the propagation of acoustic waves throughout the body of the star. The wave equation becomes

$$\frac{1}{r^2} \frac{\partial}{\partial r}\left(r^2 \frac{\partial \psi}{\partial r}\right) + \frac{1}{r^2 \sin \theta} \frac{\partial}{\partial \theta}\left(\sin \theta \frac{\partial \psi}{\partial \theta}\right) + \frac{1}{r^2 \sin^2 \theta} \frac{\partial^2 \psi}{\partial \phi^2} = -\frac{1}{c^2} \frac{\partial^2 \psi}{\partial t^2} \tag{4.5.1}$$

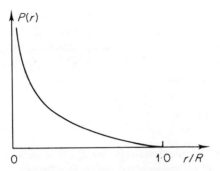

Figure 4.5.1 The radial distribution of pressure within a rotating self-gravitating star

where c is the velocity of propagation of the acoustic waves, and $\psi(r, \theta, \phi)$ represents the local fluctuation in pressure about the unexcited uniform value ψ_0. The pressure will be related to local fluctuations in density by the equation of state of the gas, and in the case of rapid density fluctuations the adiabatic relation holds:

$$p = p_0 \left(\frac{\rho}{\rho_0} \right)^{\gamma} \tag{4.5.2}$$

where p_0 and ρ_0 are the uniform values of pressure and density respectively, and $\gamma = C_p/C_v$. It follows then that $c = \sqrt{(\gamma p_0/\rho_0)}$. As we mentioned above, we might wish to more realistically incorporate gravitational effects so that $c \to c(r)$. For an ideal gas $p_0/\rho_0 = RT/M$ where M is the molecular weight of the gas and R is the gas constant, whereupon we could write $c(r) = \sqrt{(\gamma T(r)/M)}$.

If we assume a separable solution of the form

$$\psi(r, \theta, \phi, t) = R(r)\Theta(\theta)\Phi(\phi)T(t) \tag{4.5.3}$$

then we have as usual

$$T(t) = e^{-2\pi i v t} \tag{4.5.4}$$

and the separated equations

$$\frac{1}{r^2} \frac{d}{dr} \left(r^2 \frac{dR}{dr} \right) + \left(\frac{4\pi^2 v^2}{c^2} - \frac{l(l+1)}{r^2} \right) R = 0 \tag{4.5.5}$$

$$\sin \theta \frac{d}{d\theta} \left(\sin \theta \frac{d\Theta}{d\theta} \right) + [l(l+1) \sin^2 \theta - m^2]\Theta = 0 \tag{4.5.6}$$

$$\frac{d^2\Phi}{d\phi^2} + m^2\Phi = 0 \tag{4.5.7}$$

where $l(l+1), m^2$ are the usual separation constants. The solution of the angular equations follows the usual pattern subject to the condition on the angular solutions that $\Theta(\theta) = \Theta(\theta + 2\pi)$, $\Phi(\phi) = \Phi(\phi + 2\pi)$. m must therefore be an integer if $\Phi(\phi)$ is to be everywhere single valued and continuous. Equation (4.5.6) is the associated Legendre equation which has well-behaved solutions provided l is an integer and $l \geqslant m \geqslant -l$ and the general angular solution is given in terms of the surface harmonic functions

$$Y(\theta, \phi) = \sum_{l=0}^{\infty} \sum_{m=-l}^{l} A_{lm} P_l^m(\cos \theta) \, e^{\pm im\phi} \tag{4.5.8}$$

These functions have been extensively discussed in Sections 4.1 and 4.3.

The radial equation is, however, of unfamiliar form

$$\frac{d^2 R}{dr^2} + \frac{2}{r} \frac{dR}{dr} + \left(k^2 - \frac{l(l+1)}{r^2} \right) R = 0 \tag{4.5.9}$$

where $k^2 = 4\pi^2 \nu^2 / c^2$. It is very similar to the Bessel equation (3.2.5) but for the factor 2 in the first radial derivative. The parameter $l(l+1)$ is always an integer as required in Bessel's equation, but the solutions will differ from the cylindrical functions developed in the earlier sections. Provided the gradient dR/dr is not too large, we might anticipate that the solutions to equation (4.5.9) will be closely related to the Bessel functions of the first and second kinds. It turns out that we may solve equation (4.5.9) exactly by making the substitution

$$R(r) = r^{-\frac{1}{2}} \mathscr{R}(r) \tag{4.5.10}$$

whereupon equation (4.4.9) becomes

$$\frac{d^2 \mathscr{R}}{dr^2} + \frac{1}{r} \frac{d\mathscr{R}}{dr} + \left(\frac{4\pi^2 \nu^2}{c^2} - \frac{(l+\frac{1}{2})^2}{r^2} \right) \mathscr{R} = 0 \tag{4.5.11}$$

If we make the further substitution $\xi = 2\pi\nu r/c$ then we obtain from equation (4.5.11):

$$\frac{d^2 \mathscr{R}}{d\xi^2} + \frac{1}{\xi^2} \frac{d\mathscr{R}}{d\xi} + \left(1 - \frac{(l+\frac{1}{2})^2}{\xi^2} \right) \mathscr{R} = 0 \tag{4.5.12}$$

and this is *Bessel's equation of half-integral order*. The solutions to this equation are the *spherical* Bessel functions of the first and second kinds, and are written, transforming back into the original variables:

$$R(r) = r^{-1/2} \{ A_l J_{l+\frac{1}{2}}(kr) + B_l Y_{l+\frac{1}{2}}(kr) \} = \sqrt{\left| \frac{2k}{\pi} \right|} \{ A_l j_l(kr) + B_l y_l(kr) \} \tag{4.5.13}$$

which represents the radial solution to the present problem. $j_l(kr)$ are termed the spherical Bessel functions of order l, and of the first kind. The functions $y_l(kr) = j_{-l}(kr)$ are the spherical Bessel functions of the second kind, and the first few such functions are shown in Figure 4.5.2, and their analytic expression

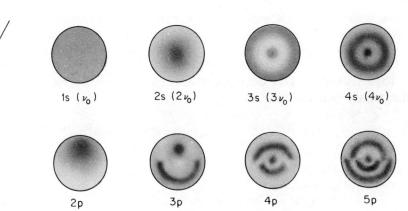

Figure 4.5.2 The first few normal radial modes of vibration within a rotating gaseous star. The spherically symmetric s modes, corresponding to $l = 0$ have the radial distributions $j_0(kr)$. The p modes ($l = 1$) have the radial distributions $j_1(kr)$, and so on. Only the zero-order spherical Bessel function has a central maximum—all the others have a maximum displaced from the centre

in Table 4.5.1. The solution of equation (4.5.12) and the mathematical properties of these functions are discussed in Appendix III. As we see, the solutions of the second kind diverge at the origin and are not therefore usually retained, except in exterior problems. In the present case we may set $B_l = 0$.

At the surface of the star we assume the Dirichlet boundary condition for the pressure $\psi(r = a) = 0$, in which case we immediately obtain from equation (4.5.13) the condition

$$R(a) = J_{l+\frac{1}{2}}(ka)/a^{\frac{1}{2}} = \sqrt{\left(\frac{2k}{\pi}\right)} j_l(ka) = 0 \qquad (4.5.14)$$

For stationary modes of vibration the nodes of this function must therefore coincide with the surface of the star, and this establishes the discrete set of frequencies sustained by the system. From the functional form of $j_0(k_n a)$ (Table 4.5.1) we might anticipate that the normal mode frequencies are harmonic, and indeed $v_{n0} = nc/2a$, where n is an integer and represents the nth crossing of the

Table 4.5.1. Spherical Bessel functions of the first and second kinds

Order	$j_l(x)$	$y_l(x) = -j_{-(l+1)}(x)$
$l = 0$	$\dfrac{\sin x}{x}$	$-\dfrac{\cos x}{x}$
$l = 1$	$\dfrac{\sin x}{x^2} - \dfrac{\cos x}{x}$	$-\dfrac{\cos x}{x^2} - \dfrac{\sin x}{x}$
$l = 2$	$\left(\dfrac{3}{x^3} - \dfrac{1}{x}\right)\sin x - \dfrac{3}{x^2}\cos x$	$\left(-\dfrac{3}{x^3} + \dfrac{1}{x}\right)\cos x - \dfrac{3}{x^2}\sin x$

radial axis. Subsequent orders are not harmonic however and the normal frequencies for the radial distribution proportional to $j_1(ka)$ are $4.4934c/2\pi a$, $7.72c/2\pi a$, $10.89c/2\pi a$, $14.08c/2\pi a, \ldots$, and for $j_2(ka)$ the frequencies occur at $5.77c/2\pi a$, $9.11c/2\pi a$, $12.34c/2\pi a$, $15.52c/2\pi a, \ldots$.

The eigenfunctions for the pressure distribution in the spherical interior of the star are therefore of the form

$$\psi_n(r, \theta, \phi, t) = \sum_{l=0}^{\infty} \sum_{m=-l}^{l} A_{lm} P_l^m(\cos\theta)\, e^{\pm im\phi} j_l(k_n r)\, e^{-2\pi i v t} \qquad (4.5.15)$$

and some of the lower-order pressure distributions are shown in longitudinal sections in Figure 4.5.2.

It is clear from the subscripts on the frequency v_{nl} that for a given pair of values l and n the parameter m may adopt all values $0, 1, 2, \ldots, l$, and the degeneracy is therefore $(2l + 1)$-fold. The standing waves for $l, m \ll n$ approximate plane waves and are reflected almost normally from the boundary of the star. These tend to focus strongly at the centre of the sphere. The first radial maximum in the distribution occurs at approximately $m/k \sim ma/\pi n$, and thereafter the amplitude of the pressure fluctuations falls off approximately inversely with radial distance. Conversely, waves for which l or m is much larger than n tend to avoid the centre of the sphere.

Many of the normal modes have very similar frequencies, and therefore hybridize extensively, even in the absence of a perturbation. Thus the frequencies of the 5s and 6d modes for example differ by only one part in one hundred, and if these modes are excited together then the 5s–6d hybrid will behave as if it were a normal vibrational mode for about 20 cycles, after which the two components move out of phase. Similarly the 3s–4d hybrid will act as a normal mode for about 6 cycles, the frequency differing by about 1 part in 30.

Suppose now that the sound velocity becomes a purely radial function. The original wave equation (4.5.1) will still separate provided c is independent of θ and ϕ. The analysis follows through as before except that now (cf. equation (4.5.13)), approximately,

$$R(r) = J_{l+\frac{1}{2}}(k(r)r)/r^{\frac{1}{2}} = \sqrt{\left(\frac{2k(r)}{\pi}\right)} j_l(k(r)r) \qquad (4.5.16)$$

where $k(r) = 2\pi v/c(r)$. If we assume some functional form for $k(r)$ (or equivalently, $c(r)$) then the spherical Bessel functions are now understood to be modified by their progressive radial collapse (or extension), and this is shown schematically for $j_0(k(r)r)$ where $k(r)$, the radial scaling parameter, varies as $k_0 \exp r$, Figure 4.5.3.

The normal mode frequencies will be determined by the coincidence of the roots of $j_n(k(r)r)$ with the stellar surfaces as before, and they will be considerably different from the values indicated in Figure 4.4.1. In particular the harmonicity of the s modes will be destroyed, but the angular solutions will remain as before, as will the $(2l + 1)$-fold degeneracy.

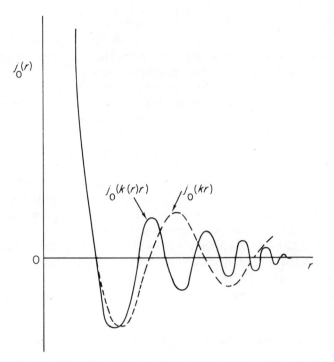

Figure 4.5.3 The radially collapsed pseudo-Bessel function
$$j_0(k(r)r)$$

Finally, we may briefly mention the solar *coronal heating problem* according to which there is some difficulty in accounting for coronal temperatures which are several orders of magnitude higher than the stellar interior. It is now generally agreed that the principal source of coronal heating must be some non-thermo-dynamic dissipative mechanism probably arising from the interior acoustic and hydrodynamic processes. The granular appearance of the sun's photosphere is thought to represent evidence of the turbulent and convective transport of energy to the surface, and at these densities non-dissipative acoustic modes such as those described in this section may be sustained. The condition for longitudinal acoustic wave propagation is that the wavelength should be much greater than the atomic mean free path. Beyond the chromosphere, however, this condition is no longer satisfied and the acoustic wave is thought to initiate a dissipative *shock wave*.

Since the velocity of propagation of a wave depends explicitly upon what one might loosely call a 'rigidity' or 'stiffness constant', it follows that the propagation velocity in a medium of zero 'rigidity' will be zero—a tensionless string, for example. On this basis a plasma cannot sustain the propagation of transverse waves since this merely involves the parallel displacement of sheets of charge: no restoring force develops. Longitudinal waves can propagate in a plasma, however, precisely because of the development of compressions of charge density. In the presence of external magnetic fields, on the other hand, the plasma develops a non-zero 'rigidity' with regard to the propagation of transverse, or *Alfvén waves*. The velocity of propagation will increase with magnetic field strength, and because the force exerted on a moving charge in a magnetic field is at right angles to the direction of motion, we expect the waves to be circularly polarized. Alfvén has attempted to explain the development of sunspots on the solar photosphere in terms of the outbreak of stable magneto-hydrodynamic vortices at the sun's surface. Many of the observed characteristics of sunspots such as radial and circulatory motions of the plasma in the sunspot (Evershed–Abetti effect) and Zeeman splitting of the spectral lines are consistent with the propagation of Alfvén waves from within the stellar interior. The origin of the intense magnetic fields required before Alfvén waves will propagate is not altogether clear: it is generally assumed that at the stellar interior there will be a convective zone and the Ampèrean currents associated with the convective motions of the plasma will generate such a field. Nevertheless, it does appear that the dissipation of Alfvén waves in the coronal region would provide a mechanism for the transport of large amounts of energy from the stellar interior out beyond the chromosphere. Such coronal heating processes are, at this stage however, purely speculative.

4.6 Partial Wave Analysis and Elastic Scattering

We now consider the elastic scattering of an incident plane wave by a spherically symmetric scattering potential $V(r)$. In the original analysis due to Rayleigh the partial wave analysis which we shall present here was developed for the scattering of acoustic waves by a spherical object. The advent of quantum mechanics has resulted in the extension to the scattering of de Broglie waves, and we shall consider in particular the elastic potential scattering of an incident beam of particles.

Central to the analysis is the spherical polar representation of a plane wave, since the incident parallel beam of particles and the scattered wave front are essentially planar at distances large with respect to the de Broglie wavelength and the range of the scattering potential. This, of course, is just the experimental situation in a scattering experiment. A principal object of such an experiment is to determine the nature of the scattering potential $V(r)$ from the scattering amplitude $f(\theta)$. In particular we may experimentally determine the *differential cross-section* for elastic scattering as the scattered intensity into the solid angle $d\Omega$, thus

$$d\sigma = |f(\theta)|^2 \, d\Omega \qquad (4.6.1)$$

for unit incident intensity. Such a quantity may be determined, for example, from the distribution of scattered intensity on a photographic plate in an X-ray or α-particle scattering experiment. The integrated quantity, the *total cross-section*, is simply given as

$$\sigma = \int |f(\theta)|^2 \, d\Omega \qquad (4.6.2)$$

The incident beam of particles e^{ikz} travelling along the positive z-axis has the associated de Broglie wavenumber k which it regains after the elastic scattering process. The particles move in a region of constant potential and so Schrödinger's equation reduces to the elementary wave equation $\nabla^2\psi + k^2\psi = 0$. Assuming the general representation of the incident wave in spherical coordinates:

$$e^{ikz} = e^{ikr\cos\theta} = R(r)\Theta(\theta)\Phi(\phi) \tag{4.6.3}$$

we obtain the usual separated equations

$$\frac{1}{r^2}\frac{d}{dr}\left(r^2\frac{dR}{dr}\right) + \left\{k^2 - \frac{l(l+1)}{r^2}\right\}R = 0 \tag{4.6.4}$$

$$\frac{1}{\Theta\sin\theta}\frac{d}{d\theta}\left(\sin\theta\frac{d\Theta}{d\theta}\right) + l(l+1) - \frac{m^2}{\sin^2\theta} = 0 \tag{4.6.5}$$

$$\frac{d^2\Phi}{d\phi^2} + m^2\Phi = 0 \tag{4.6.6}$$

Obviously for spherically symmetric scattering potentials we expect axially symmetric scattering distributions, and we may therefore set $m = 0$ throughout the analysis. Equation (4.6.5) is the Legendre equation having unassociated solutions of the form $P_l^0(\cos\theta)$, and equation (4.6.4) has solutions in terms of the spherical Bessel functions of the first kind, $j_l(kr)$: those of the second kind are infinite at the origin and must be rejected. The plane wave may therefore be represented in terms of the axially symmetric series

$$e^{ikz} = \sum_{l=0}^{\infty} A_l P_l^0(\cos\theta)j_l(kr) \tag{4.6.7}$$

where the coefficients A_l remain to be determined. Equation (4.6.7) is the partial wave representation of the plane wave in terms of the unassociated spherical harmonics.

We may associate an angular momentum with each of the partial waves about the scattering centre just as we can in the classical case in terms of the particle velocity and impact parameter. Classically, of course, any of a continuum of values may be adopted whilst quantum mechanically the particle moving in the scattering field is restricted to certain quantized values of the angular momentum. This may be visualized in the case of a parallel incident beam of particles as a series of axially symmetric cylindrical zones as shown in Figure 4.6.1. The inner radii of the zones is given as $nl\lambda/2\pi, n = 0, 1, 2, \ldots$, and the first and largest maximum of $j_l(kr)$ occurs at $kr = 2\pi r/\lambda \sim 1\cdot5l$. Thus the lth component of the partial wave has such a radial distribution of particles that they are principally contained in the zone $l\lambda/2\pi$ to $(l+1)\lambda/2\pi$, and the associated angular momentum is consequently $l\lambda/2\pi$. Clearly, only for long range potentials, and very high velocity particles is anything other than s wave ($l = 0$) scattering likely to be encountered.

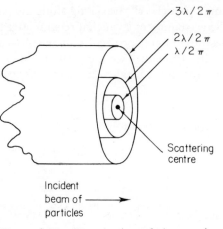

3λ/2π

2λ/2π
λ/2π

Scattering
centre

Incident
beam of ⟶
particles

Figure 4.6.1 Quantization of the angular
momentum of the incident beam of particles
about the scattering centre

The criterion for l wave scattering to dominate is simply that $l\lambda/2\pi > a$, where a is the radius of the scattering centre.

For purely s wave scattering we therefore require $2\pi a \sqrt{(2mE)}/h < 1$ where m and E are the mass and energy of the scattered particle. For the scattering of neutrons by protons, for example, we may expect pure s scattering for beam energies less than 10 MeV. An upper limit on the total *absorption* cross-section for the lth partial wave is given as the zonal area

$$\sigma_l(\text{abs}) = \pi(l+1)^2(\lambda/2\pi)^2 - \pi l^2(\lambda/2\pi) = (2l+1)\lambda^2/4\pi \qquad (4.6.8)$$

This is different to the maximum elastic *scattering* cross-section which, as we shall see is $4(2l+1)\lambda^2/4\pi$, four times greater.

In Figure 4.6.2 we compare the classical and quantum mechanical models for elastic and particle scattering by an atomic nucleus. In the quantum mechanical case the angular distribution of scattering amplitude, $f(\theta)$, is indicated by the density of the spherical wavefronts. We should expect the scattered amplitude to vary as $f(\theta)\,e^{ikr}/r$ at large distances from the scattering centre, assuming that the number of particles is conserved and that the intensity falls off as an inverse square. Under these circumstances we may write

$$\psi_{\text{total}} = \psi_{\text{incident}} + \psi_{\text{scattered}}$$

$$= e^{ikz} + f(\theta)\frac{e^{ikr}}{r} \qquad (4.6.9)$$

Now we have already seen that ψ_{incident} may be written (equation (4.6.7))

$$\psi_{\text{incident}} = e^{ikz} = \sum_{l=0}^{\infty} A_l j_l(kr) P_l^0(\cos\theta) \qquad (4.6.10)$$

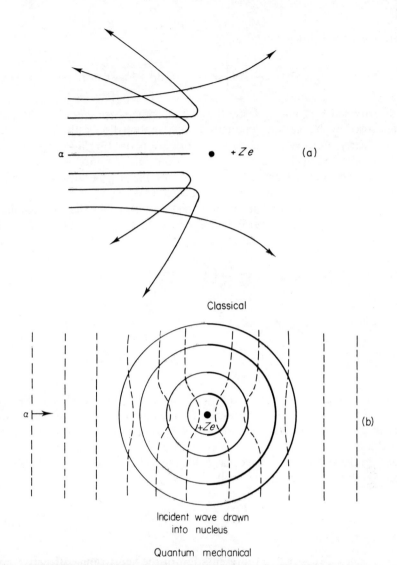

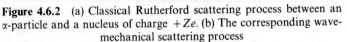

Figure 4.6.2 (a) Classical Rutherford scattering process between an α-particle and a nucleus of charge $+Ze$. (b) The corresponding wave-mechanical scattering process

where for large r, $j_l(kr)$ adopts the asymptotic form

$$j_l(kr) \rightarrow \frac{1}{kr} \sin\left(kr - \frac{l\pi}{2}\right) \quad \text{for } kr \gg l \tag{4.6.11}$$

It may be shown quite simply that the total wave is similarly given as

$$\psi_{\text{total}} = \sum_{l=0}^{\infty} A_l \, e^{i\delta_l} R_l(kr) P_l^0(\cos\theta) \tag{4.6.12}$$

where the asymptotic form

$$R_l(kr) = \frac{1}{kr} \sin\left(kr - \frac{l\pi}{2} + \delta_l\right) \quad \text{for } kr \gg l \tag{4.6.13}$$

holds provided $V(r)$ falls off faster than $1/r$. δ_l is the *phase shift* of the lth partial wave caused by the scattering potential $V(r)$; we shall discuss this in some detail later. Clearly, however, δ_l contains information on $V(r)$ although an exact analytic relationship between the two can only be obtained for a few simple forms of scattering potential. Nevertheless, $V(r)$ may generally be determined by numerical integration in more complicated cases.

Inserting the asymptotic forms for $j_l(kr)$ and $R_l(kr)$ in the expressions for ψ_{incident} and ψ_{total}, and rewriting in exponential form we obtain after some manipulation

$$\psi_{\text{scattered}} = \frac{e^{ikr}}{2ikr} \sum_{l=0}^{\infty} (2l + 1)(e^{2i\delta_l} - 1)P_l^0(\cos\theta) \tag{4.6.14}$$

i.e.

$$f(\theta) = r\,e^{-ikr}\psi_{\text{scattered}}$$

$$= \frac{1}{2ik} \sum_{l=0}^{\infty} (2l + 1)(e^{2i\delta_l} - 1)P_l^0(\cos\theta) \tag{4.6.15}$$

where, in a straightforward but tedious way, it may be shown that $A_l = (2l + 1)i^l$. Now that we have a general expression for the amplitude of the scattered wave and hence $f(\theta)$, the differential and total elastic scattering cross-sections follow immediately.

One simple example which we may calculate is elastic s wave ($l = 0$) scattering by a deep spherical rectangular well:

$$V(r) = -\varepsilon \quad r < a$$
$$= 0 \quad r > a \tag{4.6.16}$$

and this gives a reasonable representation of the short-range attractive interaction between a nucleus and a neutron for which there are no Coulombic effects. The incident wave is now given in asymptotic form as (from equations (4.6.10), (4.6.11), $l = 0$),

$$\psi_{\text{incident}} = e^{ikz} = \frac{\sin kr}{kr} = \frac{e^{ikr} - e^{-ikr}}{2ikr} \quad (l = 0) \tag{4.6.17}$$

where $k = \sqrt{(2mE)/\hbar^2}$.

Similarly the (asymptotic) total wave is

$$\psi_{\text{total}} = \frac{B \sin(kr + \delta_0)}{kr} \tag{4.6.18}$$

where B is a constant to be determined, so that

$$\psi_{\text{scattered}} = \frac{e^{ikr}}{r} f(\theta) = \psi_{\text{total}} - \psi_{\text{incident}}$$

$$(4.6.19)$$

$$= \frac{1}{2ikr} \{(B\, e^{i\delta_0} - 1)\, e^{ikr} - (B\, e^{-i\delta_0} - 1)\, e^{-ikr}\}$$

Now, e^{-ikr} has the significance of an *incoming* wave and in a scattered wave its coefficient must be zero, that is

$$B = e^{i\delta_0} \qquad (4.6.20)$$

It then immediately follows from equation (4.6.19) that

$$f(\theta) = \frac{1}{2ik}(e^{2i\delta_0} - 1) = \frac{e^{i\delta_0} \sin \delta_0}{k} \qquad (4.6.21)$$

The total elastic cross-section for s wave scattering therefore follows from equation (4.6.2) as

$$\sigma_0 = 4\pi f(\theta) f(\theta)^* = \frac{4\pi}{k^2} \sin^2 \delta_0 \qquad (4.6.22)$$

in terms of the real phase shift δ_0. σ is seen to be an isotropic (θ-independent) function for s wave scattering. From equation (4.6.22) we see that $(2 \sin \delta_0/k)$ represents the effective radius of the target, and since $\sin \delta_0 \ll 1$ we have $\sigma \leqslant 4\pi/k^2$ providing an upper geometric limit for s wave scattering at an energy corresponding to k. Moreover, this limit is independent of the scattering potential.

We have so far related the scattering cross-section to the s wave phase shift δ_0—it now remains to associate δ_0 with the scattering potential $V(r)$. This is done by matching the wavefunctions inside and outside the well such that ψ is everywhere single valued and continuous. In doing so we also see how the phase shift develops.

The total wavefunction ψ_{total} must satisfy the radial Schrödinger equation

$$\frac{d^2\psi}{dr^2} + \frac{2}{r}\frac{d\psi}{dr} - \frac{2m}{\hbar^2}(E - V(r))\psi = 0 \quad (l = 0) \qquad (4.6.23)$$

outside the well ($r > a$), whilst the wavefunction

$$\psi_{\text{inside}} = \frac{C}{r} \sin k_1 r; \quad k_1^2 = \frac{2m}{\hbar^2}(E + \varepsilon) \qquad (4.6.24)$$

being $j_0(kr)$ is seen to satisfy equation (4.6.23) inside the well ($r < a$). The parameters C and δ_0 are to be determined such that ψ and $\partial\psi/\partial r$ are single-valued and continuous at $r = a$. We therefore obtain the boundary condition

$$\psi_{\text{total}}(a) = \psi_{\text{inside}}(a): \quad \frac{e^{i\delta_0}}{k} \sin(ka + \delta_0) = C \sin k_1 a$$

$$\frac{\partial \psi_{\text{total}}(a)}{\partial r} = \frac{\partial \psi_{\text{inside}}(a)}{\partial r}: \quad e^{i\delta_0} k \cos(ka + \delta_0) = Ck_1 \cos k_1 a$$

(4.6.25)

The ratio of these two equations gives

$$\delta_0(k) = -ka + \tan^{-1}\left(\frac{k}{k_1} \tan k_1 a\right)$$

(4.6.26)

where

$$k^2 = \frac{2\pi m E}{\hbar^2}, \quad k_1^2 = \frac{2m(E + \varepsilon)}{\hbar^2}$$

We could, therefore, express σ directly in terms of the parameters of the scattering potential $V(r)$. We should note, however, that the phase shift δ_0 is indefinite by an integral multiple of π, because of the lack of uniqueness in the solution of equation (4.6.26). Indeed, for certain values of k the phase shift will be *zero*. This corresponds physically to the potential hole being just strong enough to introduce one or more additional complete wavelengths within its range of action. The wave outside the well remains unaffected—the gain of one or two more wavelengths being unobservable at infinity.

In the limit of low particle energies ($k \to 0$) we see from equation (4.6.26) that $\delta_0 \to 0$ and so

$$f(\theta) \to \frac{\delta_0}{k} \quad \text{as } k \to 0$$

i.e.

(4.6.27)

$$\sigma_0 = 4\pi f(\theta) f(\theta)^* = \frac{4\pi \delta_0^2}{k^2}$$

If the elastic s wave scattering partial cross-section σ_0 passes through zero (with δ_0) for some value of k, whilst all the other cross-sections σ_l for higher partial wave scattering still remain small, then the total scattering cross-section will show a deep minimum at that velocity, and this would be expected to occur for attractive fields in general. Indeed argon becomes abnormally 'transparent' to low energy (<10 eV) electrons, whilst the scattering increases at slightly higher energies. This has been widely observed, and is termed the *Ramsauer effect*. δ_0/k has the dimension of length and $(-\delta_0/k) = d$ is termed the *scattering length*. For certain values of the incident energy near the discrete stationary levels within the well corresponding to the bound states, strong absorption can occur. For this *resonance scattering* d is likely to possess a large imaginary component.

We may consider the scattering length a little further by considering the total wavefunction in the limit $k \to 0$. From equations (4.6.18) and (4.6.20) we have

$$\psi_{\text{total}} = \frac{e^{i\delta_0}}{kr} \sin(kr + \delta_0)$$

$$= 1 + \frac{\delta_0}{kr} \quad \text{as } k \to 0$$

i.e.

$$r\psi_{\text{total}}(r) = r - d \quad \text{as } k \to 0 \tag{4.6.28}$$

so that the scattering length has the physical significance of being the zero energy radial intercept of the function $r\psi_{\text{total}}$. For this reason the scattering length is sometimes known as the *Fermi intercept* of $r\psi_{\text{total}}(r)$. d is positive or negative according as to whether $k = 2\pi mE/\hbar^2$ relates to a bound $(E < 0)$ or unbound $(E > 0)$ state. Thus the scattering length is positive for scattering from a bound state and negative for scattering from an unbound state.

We may finally observe that in the limit $k \to 0$ the zero energy total scattering cross-section becomes

$$\sigma = 4\pi|f(\theta)|^2 = 4\pi d^2 \tag{4.6.29}$$

from equation (4.11.27), which is the same as the zero energy scattering from an impenetrable sphere of radius d. This is, however, four times larger than the classical cross-section for point particles off a hard sphere of the same radius.

The phase shift of an s wave on a rectangular well may be simply understood as a consequence of the continuity of the internal and external wavefunctions at $r = a$. Thus, in Figure 4.6.3 we see from a comparison of the external wavefunction ψ_{total} in the absence and presence of the scattering potential that the external wave is drawn in or phase shifted by δ_0 relative to the unscattered wave. If the potential were repulsive the phase shift would be negative and the scattered total wave 'pushed-out', again as a consequence of the boundary conditions.

Perhaps the origin of the phase shift may be made more physically apparent by recalling that the velocity of propagation of matter waves depends upon the local potential. Thus, a 'refractive index' entirely analogous to its optical counterpart may be defined as

$$\mu = \frac{\text{wave velocity in free space}}{\text{wave velocity in scattering field}}$$

Indeed, it is not very difficult to show that

$$\mu = \sqrt{\frac{E - \varepsilon}{E - \varepsilon_0}}$$

Figure 4.6.4 Nature of the scattering potential for s wave scattering $(l = 0)$ and higher partial wave $(l > 0)$ scattering.

156

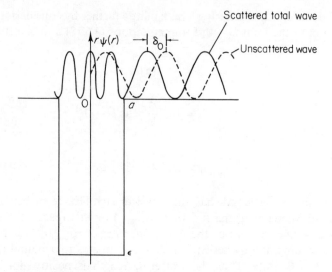

Figure 4.6.3 Development of the phase shift δ between an
unscattered and the total scattered wave

where E is the total energy of the matter wave and ε represents the local potential, and will generally be a function of position. ε_0 is the potential outside the scattering field. If ε is a negative quantity, so that μ is a real, positive number > 1, then from the first μ relation we conclude that the particle tends to 'hang around' in the scattering field rather longer than we might otherwise have expected. The subsequently scattered wave is in consequence phase shifted by the quantity δ_l. Whilst this is no more than a hand-waving argument, it does help identify the physical origin of the shift.

The more complicated situation in which we have both attractive and repulsive regions applies, of course, to Rutherford scattering in which the repulsive interaction is Coulombic. We are generally concerned with particle energies which are insufficient to penetrate the 'nuclear' region, although at sufficiently high energies *anomalous scattering* may occur in which coherent scattering both from the discontinuity of potential at the nuclear surface and from the purely Coulombic envelope give rise to interference terms. Under these circumstances we can write the differential cross-section as

$$\mathrm{d}\sigma = |f_{\text{nuclear}}(\theta) + f_{\text{coulombic}}(\theta)|^2 \, \mathrm{d}\Omega \tag{4.6.30}$$

and in consequence there will be cross-terms in which the nuclear scattering is coupled to the purely Coulombic component.

We observed at the outset that for the partial wave analysis to be applicable the scattering potential $V(r)$ must fall off *faster* than $1/r$: clearly this does not apply in the case of Rutherford scattering for which $V(r) = Zze^2/r$, where Ze is the effective nuclear charge and ze is the charge on the incident particle.

Under these circumstances the asymptotic expressions for the s wave phase shift δ_0 do not apply. Nevertheless, Gordon has been able to show that for inverse square force fields only, the scattering distribution is given as

$$f(\theta) = \frac{Zze^2}{2m_e v} \frac{1}{\sin^2(\theta/2)} \qquad (4.6.31)$$

which leads to the classical Rutherford value for the differential scattering cross-section. Williams has pointed out on the basis of a dimensional generalization that if the interparticle force varies as r^n, then the scattering cross-section will vary as h^{4+2n}, where h is Planck's constant. Obviously for inverse square fields only is the cross-section independent of the quantum of action and the classical and quantum mechanical results likely to coincide.

We have so far restricted ourselves to relatively low-energy s wave scattering: at higher energies we must account for scattering modes for $l > 0$. Here we shall make no more than a passing observation on the effective modification of the scattering potential for p, d, f, ... partial wave scattering.

The radial wave equation will have the form (4.6.4)

$$\frac{d^2R}{dr^2} + \frac{2}{r}\frac{dR}{dr} + \frac{2m_e}{\hbar^2}\left(E - V(r) - \frac{\hbar^2}{2m_e}\frac{l(l+1)}{r^2}\right)R = 0 \qquad (4.6.32)$$

For s wave scattering ($l = 0$) the scattering potential is simply $V(r)$. For $l > 0$, however, the scattering potential is *effectively* $-\{V(r) + l(l+1)/r^2\}$. That is, $V(r)$ is enhanced by a *centrifugal term*. In other words the centrifugal term may be incorporated as an additional repulsive component of the scattering potential. In Figure 4.6.4 we schematically illustrate the centrifugal enhancement of the Coulomb scattering potential for high-energy particles.

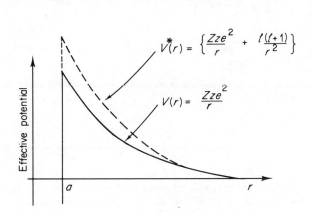

For $l > 0$ the repulsion is enhanced by a 'centrifugal' term

As a further example of a scattering problem we may consider the calculation of the quantum mechanical second virial coefficient $B_2(T)$ which arises in the density expansion of the equation of state for a dilute assembly of particles:

$$\frac{PV}{NkT} = 1 + B_2(T)\rho + B_3(T)\rho^2 + \cdots \qquad (4.6.33)$$

Physically the virial coefficients represent *cluster integrals* over configurations of $2, 3, \ldots$ interacting particles in the gas so that successive terms in the series represent the two- three-, . . . , body contributions to the equation of state and thereby account for the departure of a real system from idealism.

The classical Boltzmann expression for the second virial coefficient representing the effect of two-body interactions in the assembly is simply

$$B_2(T) = -2\pi \int_0^\infty \left[\exp\left(-\frac{\Phi(r)}{kT}\right) - 1 \right] r^2 \, dr \qquad (4.6.34)$$

where $\Phi(r)$ is the pair potential, and the success with which this describes the temperature dependence for classical systems is shown in Figure 4.6.5. For a

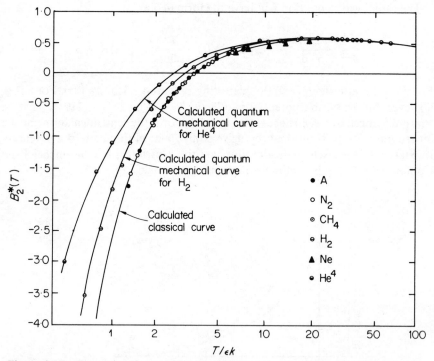

Figure 4.6.5 The reduced second virial coefficient for a number of simple systems estimated on the basis of equation (4.6.34). The agreement for classical systems is excellent: the quantal systems depart significantly at low temperatures however. The agreement with the quantum mechanical calculation (equation 4.6.36)) is very good indeed. $(B_2^*(T) = B_2(T)/(2\pi\sigma^3/3))$. (J. O. Hirschfelder, C. F. Curtiss and R. B. Bird, *Molecular Theory of Gases and Liquids*, John Wiley and Sons, Inc., 1964)

typical pair potential of Lennard–Jones form (Figure 4.6.6) equation (4.6.34) tends to zero as $T \to \infty$. At low temperatures, as $T \to 0$ and the attractive negative region of the pair potential dominates the interaction, $B_2(T)$ becomes large and negative. At these low temperatures the quantum mechanical curves are seen to depart significantly from their classical counterparts.

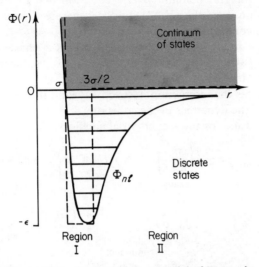

Figure 4.6.6 A typical pair potential of Lennard–Jones form showing the development of the discrete and collisional continuum of states. The degenerate discrete states are designated Φ_{nl}. The square well model interaction is also shown: this simple function provides a convenient mathematical model whilst retaining the principal qualitative features of the realistic interaction

Even for an ideal (non-interacting) assembly of particles, an *exchange* correction will be required in the quantum mechanical case simply on account of the different (i.e. non-Boltzmann) statistical class to which the assembly belongs. The stringent restriction on fermions, for example, regarding the occupation number of the energy levels means that they behave *as if* there were an additional repulsion or exclusion operating between the particles. Thus the Pauli repulsion between fermions increases the pressure above the ideal value. On the other hand the relaxation of particle distinguishability assumed in Boltzmann statistics results in an apparent attraction between bosons, and the pressure is lowered relative to an ideal classical system.

The exchange correction is

$$B_2(T) = \pm \frac{1}{2}\left(\frac{\pi\hbar^2}{mkT}\right)^{\frac{3}{2}} \tag{4.6.35}$$

where the $-(+)$ sign applies to a boson (fermion) assembly.

In the quantization of the binary interaction for a realistic system, two kinds of solution will arise: a positive *continuum* of unbound levels corresponding to high-energy collisional interactions for $kT > |\varepsilon|$, and a degenerate set of discrete levels for $kT < |\varepsilon|$ corresponding to bound states. Both the continuum and discrete states arise as solutions to the radial Schrödinger equation. The classical cluster integral defining $B_2(T)$, equation (4.6.34), will therefore be replaced by a sum over the $(2l + 1)$-fold degenerate discrete states plus an integral over the continuum of collisional states.

In relative coordinates, the incident wave suffers a phase shift in the field of the scattering particle: the wave is drawn in or expelled from the vicinity of the target atom depending whether the negative or positive region of the potential dominates the interaction, respectively, and this of course represents a modification of the dynamics of the collision.

The final expression for the second virial coefficient is

$$B_2(T)_{\text{quantum}} = \underbrace{\pm \frac{1}{2}\left(\frac{\pi \hbar^2}{mkT}\right)^{\frac{3}{2}}}_{\text{exchange}} - 8\sum_l (2l + 1)\Bigg\{\underbrace{\sum_n \exp\left(-\Phi_{nl}/kT\right)}_{\text{discrete}}$$

$$+ \underbrace{\frac{1}{\pi}\int_0^\infty \frac{d\delta_l}{dp}\exp\left(-p^2/mkT\right)dp}_{\substack{\text{collisional}\\ \text{continuum}}}\Bigg\}$$

(4.6.36)

The exchange term arises even for non-interacting systems, whilst the sum over the $(2l + 1)$-fold degenerate discrete spectrum and the collisional continuum arises only for interacting assemblies. δ_l is the lth partial wave phase shift and its variation with particle momentum must be determined together with the discrete energy eigenvalues before the second virial coefficient can be determined. At low temperatures and for short range potentials the scattering is predominantly s *wave* ($l = 0$), and in order to make calculations of the second virial coefficient we need to determine δ_0. It is not generally possible to obtain analytic expressions for the phase shifts in terms of the pair potential. However, solutions in closed form may be obtained for hard spheres (for all l) and for the square well potential ($l = 0$).

The s wave phase shift for a square well interaction may be easily determined and on the basis of this we may gain some understanding of the phase shift contributions to the equation of state of more realistic systems. For this potential we wish to find a solution to the radial wave equation which describes the relative motion of the two particles with relative energy E and with $l = 0$. Since the wavefunction is to vanish at $r = \sigma$ we have (see Figure 4.6.6)

$$rR(r)_1 = A_1 \sin k_1(r - \sigma) \qquad \sigma < r < 3\sigma/2$$

$$rR(r) = A \sin (kr + \delta_0) \qquad r > 3\sigma/2$$

where

(4.6.37)

$$k_1^2 = 2m(E + \varepsilon)/\hbar^2, \quad k^2 = 2mE/\hbar^2$$

(m = reduced mass). The requirement that $rR(r)$ and $d(rR(r))/dr$ be continuous at $r = 3\sigma/2$ gives

$$\delta_0(k) = \tan^{-1}\left\{\frac{k}{k_1}\tan\frac{k_1\sigma}{2}\right\} - \frac{3}{2}k\sigma \qquad (4.6.38)$$

and the phase shifts as functions of $k\sigma$ are given graphically in Figure 4.6.7 as functions of the square well depth (cf. equation (4.6.26)).

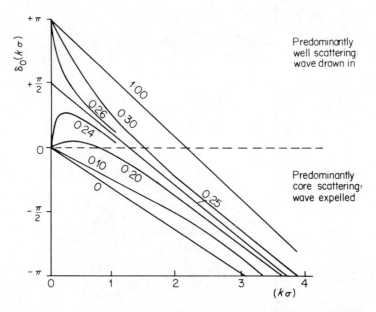

Figure 4.6.7 The s wave square well phase shift δ_0 as a function of well depth ε. At low energies δ_0 is positive corresponding to the wave being 'drawn in' to satisfy the boundary conditions at $r = 3\sigma/2$. At higher energies, and at all energies for the rigid sphere interaction ($\varepsilon = 0$) the core of the potential dominates the collision and the wave is expelled to fit the boundary condition at $r = \sigma$. At certain energies a Ramsauer effect occurs and the phase shift is zero. (J. O. Hirschfelder, C. F. Curtiss and R. B. Bird, *Molecular Theory of Gases and Liquids*, John Wiley and Sons, Inc., 1964)

We see that the square well phase shift varies discontinuously as the well depth is increased. For $\varepsilon = 0$ we have rigid sphere behaviour. As ε increases discontinuous jumps in the phase shift at $k\sigma = 0$ occur, and these correspond to the development of a discrete bound level just at the brim of the well. We see that for low-energy collisions the inward phase shift becomes more and more pronounced with increasing ε as the particle feels the dip in the potential. At high energies the dip is unnoticed and the rigid sphere phase shifts are recovered.

From the fermion curve for He³ there is a greater Pauli exclusion of the incident wave relative to the boson He⁴ phase shift (Figure (4.6.8)). From the low-energy He³ phase shift ($\delta_0 = 0$) it may be concluded that no discrete bound

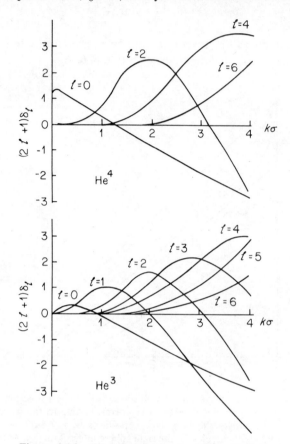

Figure 4.6.8 The phase shifts for He³ and He⁴. For all partial waves the He³ shifts exhibit a more repulsive aspect than He⁴ as we might expect for interacting fermions. At very low energies He⁴ appears to suggest there is a stable discrete bound state, whilst for He³ this appears not to be the case. Again, at low energies the attractive features of the interaction dominate and the phase shifts are positive, whilst the features of the core potential become more evident at high energies.

For higher partial wave scattering ($l > 0$) it is clear from the He⁴ curves that the centrifugal potential $l(l + 1)/r^2$ eliminates any possibility of any discrete (bound) contributions to the second virial coefficient. (J. O. Hirschfelder, C. F. Curtiss and R. B. Bird, *Molecular Theory of Gases and Liquids*, John Wiley and Sons, Inc., 1964) states exist, whilst it does appear that there is a discrete state for He^4. At high energies the phase shifts become negative and rigid-sphere *like*—the Lennard–Jones potential allows for a certain penetration at high energies.

$B_2(T)$ calculated on the basis of equation (4.6.36) is shown in Figure 4.6.5 and the agreement is seen to be excellent. A comparison of the very low temperature calculations for He^3 and He^4 are shown in Figure 4.6.9. Except at $T \sim 0\,K$,

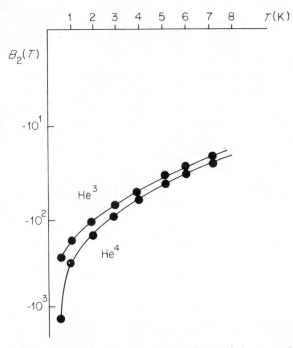

Figure 4.6.9 The low-temperature form of the second virial coefficient for He^3 and He^4

exchange effects are negligible in comparison with other terms in equation (4.6.36) thereby vindicating our use of Boltzmann statistics at all but the lowest temperatures. The difference in the $B_2(T)$ curves for He^3 and He^4 may be attributed almost entirely to the absence of discrete contributions in the former case, and the slight difference in the low-energy phase shift gradients (Figure 4.6.8).

4.7 The Hydrogen Atom

A great many quantum mechanical problems involve either a central force field, or a field which is very nearly central, and it is instructive to consider the few cases for which an exact solution may be obtained. Perhaps the two most important potentials in this respect are those of the harmonic oscillator, already discussed for the linear case (Section 2.7), and the Coulombic potential, and on the basis of these exact solutions perturbation calculations may be developed in problems for which there is no exact treatment. Here we shall begin with the solution of the Schrödinger equation for hydrogen and the hydrogenic series—He^+, Li^{++}, Be^{+++},....

The Schrödinger equation for a bound two-body (hydrogenic) central force system is

$$\nabla^2\psi + \frac{2m_e}{\hbar^2}\left(E + \frac{Ze^2}{r}\right) = 0 \qquad (4.7.1)$$

where we have written the total energy explicitly as a negative quantity, and Ze^2/r is the Coulombic interaction between the electron and the nuclear charge, Ze. We have tacitly assumed here that the nuclear mass is infinite and does not move; strictly we should replace the electron mass m_e by its *reduced* value $\mu = m_e M/(m_e + M)$ where M is the nuclear mass. Since $M \gg m_e, \mu \sim m_e$ however. Expressing equation (4.7.1) in spherical coordinates we obtain

$$\frac{1}{r^2}\frac{\partial}{\partial r}\left(r^2\frac{\partial\psi}{\partial r}\right) + \frac{1}{r^2\sin\theta}\frac{\partial}{\partial\theta}\left(\sin\theta\frac{\partial\psi}{\partial\theta}\right) + \frac{1}{r^2\sin^2\theta}\frac{\partial^2\psi}{\partial\phi^2}$$
$$+ \frac{2m_e}{\hbar^2}\left(E + \frac{Ze^2}{r}\right)\psi = 0 \qquad (4.7.2)$$

Assuming a solution of the form $\psi(r, \theta, \phi) = R(r)\Theta(\theta)\Phi(\phi)$, separation of variables yields the radial and angular equations:

$$\frac{1}{r^2}\frac{d}{dr}\left(r^2\frac{dR}{dr}\right) + \left(\frac{2m_e E}{\hbar^2} + \frac{2m_e Ze^2}{\hbar^2 r} - \frac{l(l+1)}{r^2}\right)R = 0 \qquad (4.7.3)$$

$$\frac{1}{\sin\theta}\frac{\partial}{\partial\theta}\left(\sin\theta\frac{\partial Y}{\partial\theta}\right) + \frac{1}{\sin^2\theta}\frac{\partial^2 Y}{\partial\phi^2} + l(l+1)Y = 0 \qquad (4.7.4)$$

where

$$Y(\theta, \phi) = P_l^m(\cos\theta)\,e^{\pm im\phi}$$

The angular solutions are of the usual spherical surface harmonic form, and have been extensively discussed earlier (Sections 4.1, 4.3). The radial equation is, however, quite unfamiliar. It may be simplified somewhat by making the substitutions

$$\rho = 2r\sqrt{(2m_e E)}/\hbar, \quad \lambda = Zm_e e^2/\sqrt{(2m_e E)}\hbar, \quad R(r) = \mathscr{R}(\rho)$$

Equation (4.7.3) then becomes,

$$\frac{d^2\mathscr{R}}{d\rho^2} + \frac{2}{\rho}\frac{d\mathscr{R}}{d\rho} - \left\{\pm\frac{1}{4} - \frac{\lambda}{\rho} + \frac{l(l+1)}{\rho^2}\right\}\mathscr{R} = 0 \qquad \begin{array}{l} -\frac{1}{4}\text{ for } E > 0 \\[4pt] +\frac{1}{4}\text{ for } E < 0 \end{array} \qquad (4.7.5)$$

which at large ρ reduces to (for $E < 0$)

$$\frac{d^2\mathcal{R}}{d\rho^2} - \frac{1}{4}\mathcal{R} \sim 0 \qquad (4.7.6)$$

and this has a solution of the form

$$\mathcal{R} = A \exp\left(-\frac{\rho}{2}\right) + B \exp\left(\frac{+\rho}{2}\right) \qquad (4.7.7)$$

If we are to have finite solutions at infinity then clearly we must set $B = 0$. Equation (4.7.7) then represents the asymptotic form of the radial equation. At intermediate ρ we therefore look for solutions of the form

$$\mathcal{R}(\rho) = AK(\rho)\,e^{-\rho/2} \qquad (4.7.8)$$

Substitution of equation (4.7.8) into equation (4.6.5) yields the differential equation for $K(\rho)$:

$$\rho^2\frac{d^2K}{d\rho^2} + \rho(2 - \rho)\frac{dK}{d\rho} + \{(\lambda - 1)\rho - l(l + 1)\}K = 0 \qquad (4.7.9)$$

All we can say about $K(\rho)$ so far is that it must increase less rapidly than $e^{\rho/2}$ as $\rho \to \infty$ if the solution is to remain finite at infinity, whilst at $\rho = 0$ it follows directly from equation (4.7.9) that

$$K(0)l(l + 1) = 0$$

and so for $l \neq 0$, $K(0) = 0$. A series solution of equation (4.7.9) (Appendix VII) shows that solutions finite at infinity are obtained for *positive integer values of λ only*. Moreover, the allowed values of λ are restricted such that

$$\lambda = n = l + 1, l + 2, l + 3, \ldots \qquad (4.7.10)$$

We have therefore already established the quantization of the energy eigenvalues, since from the initial substitutions we now have

$$E_n = -\frac{Z^2 e^4 m_e}{2n^2 \hbar^2} \qquad (4.7.11)$$

where n is given by equation (4.7.10). The above expression for the system of energy levels is, of course, identical to the semi-classical result of Bohr.

The function $K(\rho)$ is eventually determined to have the analytic form

$$K(\rho) = \rho^l L_{n+l}^{2l+1}(\rho) \qquad (4.7.12)$$

(see Appendix VII) where $L_r^s(\rho)$ is the associated Laguerre function of order s and degree $(r - s)$. The wavefunctions of the hydrogen atom therefore have the form

$$\psi_{nlm}(r, \theta, \phi) = A\rho^l\,e^{-\rho/2}L_{n+l}^{2l+1}(\rho)Y_l^m(\theta, \phi) \qquad (4.7.13)$$

where $\rho = 2Ze^2 m_e r/n\hbar^2$ (see equation (4.7.11)), and A is an arbitrary constant to be determined.

Notice that the energy is quantized (equation (4.7.11)) and the integer quantum numbers n, l, m are established without any boundary conditions other than the condition that the wavefunction should be everywhere single valued, continuous and finite. We see from the condition (4.7.10) that $l \leqslant n - 1$, and as usual $|m| \leqslant l$ so that in general the degeneracy for a given level E_n is

$$\sum_{l=0}^{n-1} (2l + 1) = n^2$$

These n^2 particular solutions of the same energy are of course orthogonal since the spherical harmonics $Y_l^m(\theta, \phi)$ are mutually orthogonal. We shall discuss the extensive degeneracy which develops in atomic systems, and their experimental investigation, in Section 4.10. However, it is appropriate at this stage to mention that the degeneracy with respect to l is a peculiar feature of the Coulomb interaction: in the general central force problem the energy levels are non-degenerate with respect to l. We may at this point make contact with the Sommerfeld model of the atom in which elliptic orbits of the same E but different l correspond to orbits of same energy but different angular momenta. As we shall see in Section 4.10, solutions of given (n, l) but different $|m|$ correspond to different orientations of the elliptic orbits when spatial quantization is introduced as, for example, in the Stark and Zeeman effects.

If we take into account electron spin the degeneracy of the state becomes $2n^2$, and if the spin of the nucleus is also included the degeneracy of the nth energy level increases to $2n^2(2I + 1)$ where I is the quantum number associated with the nuclear spin.

The significance of normalization of the wavefunctions and the evaluation of the constant A in equation (4.7.13) should be mentioned. In general, according to the statistical interpretation, $|\Psi(\mathbf{r}, t)|^2$ represents the probability of finding the associated particle in the volume element $d\tau$ located at $\mathbf{r}$ at time t. Since the particle is evidently *somewhere* in space, integration over the whole accessible volume must yield

$$\int_\tau |\Psi(\mathbf{r}, t)|^2 \, d\tau = 1 \tag{4.7.14}$$

that is, the particle must certainly be found somewhere within the accessible volume τ. Taking the unnormalized wavefunction ψ_{nlm} (equation (4.7.13)) we may therefore determine the coefficient A as

$$A = \frac{1}{\sqrt{\int |\psi_{nlm}(r, \theta, \phi)|^2 \, d\tau}} \tag{4.7.15}$$

where A is now termed the normalization constant. Other than in qualitative discussion it is essential to use normalized wavefunctions, although we should emphasize that not all wavefunctions may be normalized. For example, a restriction is placed on the wavefunction such that it must approach zero sufficiently rapidly at infinity for the integral in equation (4.7.15) to converge to a finite value. Thus the boundary condition at infinity is imposed on all physically acceptable wavefunctions and this, incidentally, establishes the energy quantization. If such a condition is not met the wavefunction cannot represent a finite probability distribution, and must be rejected. Again, singularities in the wavefunction may lead to difficulties in normalization. If, for example, the function has a pole at the origin of the form

$$\psi \sim \frac{1}{r^n}$$

then we have in the vicinity of the origin

$$\int |\psi|^2 \, d\tau \sim \int \frac{1}{r^{2n}} r^2 \, dr \sim \frac{1}{r^{2n-3}}$$

Clearly, for finite probability distributions singularities with $n < \frac{3}{2}$ may be allowed.

The normalization constant A in equation (4.7.13) may either be evaluated according to equation (4.7.15) or by separately normalizing the $R_{n,l}(r)$, $\Theta_l(\theta)$ and $\Phi_m(\phi)$ solutions and then forming the normalized product $\psi_{nlm} = R_{nl}(r) Y_l^m(\theta, \phi)$. We shall follow the latter course here simply so that we can demonstrate the stages of normalization. This approach also has the advantage that the normalized distributions $\Phi\Phi^*$, $\Theta\Theta^*$ and RR^* then represent the separate angular and radial distributions.

Thus, from the conditions

$$\left. \begin{array}{l} \displaystyle\int_0^{2\pi} \Phi\Phi^* \, d\phi = 1 \rightarrow A_\phi = \frac{1}{\sqrt{(2\pi)}} \\[4mm] \displaystyle\int_0^\pi \Theta\Theta^* \sin\theta \, d\theta = 1 \rightarrow A_\theta = \sqrt{\frac{(2l+1)(l-m)!}{2(l+m)!}} \end{array} \right\} \qquad (4.7.16)$$

and

$$\int_0^\infty RR^* r^2 \, dr = 1 \rightarrow A_r = \sqrt{\frac{4(n-l-1)!}{a^3 n^4 [(n+l)!]^3}}$$

the total normalization constant appearing in equation (4.7.13) is then compounded as $A = A_r A_\theta A_\phi$. If we now take $\rho = 2Ze^2 m_e r/n\hbar^2$, $a = \hbar^2/Ze^2 m_e$ the radius of the first quantized Bohr orbit, then the first few orthonormal hydrogen wavefunctions are given in Table 4.7.1, some of which are shown in Figure 4.7.1.

Various attempts were made by Pauli, Heisenberg and Jordan to incorporate electron spin and relativity corrections into the Schrödinger equation. Dirac, starting with a relativistic wave equation, found that the properties of the spinning electron emerged automatically, and moreover, in the case of hydrogen Dirac's energy levels coincided with the relativistic correction proposed by Sommerfeld. The coincidence is accounted for in terms of the degeneracy of the Dirac levels. The energy associated with an eigenstate of given principal quantum number n depends only upon the *total* angular momentum vector j, and

Table 4.7.1. Orthonormalized eigenfunctions of the hydrogen
atom

$$\rho = 2Z\,e^2 m_e r/n\hbar^2 \,; a = \hbar^2/Z\,e^2 m_e$$

ψ_{nlm}	
ψ_{100}	$\dfrac{1}{\sqrt{\pi}}\dfrac{1}{a^{\frac{3}{2}}}\,e^{-\rho/2}$
ψ_{200}	$\dfrac{1}{4\sqrt{(2\pi)}}\dfrac{1}{a^{\frac{3}{2}}}\left(2 - \dfrac{\rho}{2}\right)e^{-\rho/2}$
ψ_{211}	$\dfrac{1}{8\sqrt{(2\pi)}}\dfrac{1}{a^{\frac{3}{2}}}\dfrac{\rho}{2}\,e^{-\rho/2}\sin\theta\,e^{i\phi}$
ψ_{210}	$\dfrac{1}{4\sqrt{(2\pi)}}\dfrac{1}{a^{\frac{3}{2}}}\dfrac{\rho}{2}\,e^{-\rho/2}\cos\theta$
ψ_{21-1}	$\dfrac{1}{8\sqrt{\pi}}\dfrac{1}{a^{\frac{3}{2}}}\dfrac{\rho}{2}\,e^{-\rho/2}\sin\theta\,e^{-i\phi}$
ψ_{300}	$\dfrac{1}{81\sqrt{(3\pi)}}\dfrac{1}{a^{\frac{3}{2}}}\left(27 - 18\dfrac{\rho}{2} + 2\dfrac{\rho^2}{4}\right)e^{-\rho/2}$
ψ_{311}	$\dfrac{1}{81\sqrt{\pi}}\dfrac{1}{a^{\frac{3}{2}}}\left(6 - \dfrac{\rho}{2}\right)\dfrac{\rho}{2}\,e^{-\rho/2}\sin\theta\,e^{i\phi}$
ψ_{310}	$\dfrac{\sqrt{2}}{81\sqrt{\pi}}\dfrac{1}{a^{\frac{3}{2}}}\left(6 - \dfrac{\rho}{2}\right)\dfrac{\rho}{2}\,e^{-\rho/2}\cos\theta$
ψ_{31-1}	$\dfrac{1}{81\sqrt{\pi}}\dfrac{1}{a^{\frac{3}{2}}}\left(6 - \dfrac{\rho}{2}\right)\dfrac{\rho}{2}\,e^{-\rho/2}\sin\theta\,e^{-i\phi}$
ψ_{322}	$\dfrac{1}{81\sqrt{(4\pi)}}\dfrac{1}{a^{\frac{3}{2}}}\dfrac{\rho^2}{4}\,e^{-\rho/2}\sin^2\theta\,e^{i2\phi}$
ψ_{321}	$\dfrac{1}{81\sqrt{\pi}}\dfrac{1}{a^{\frac{3}{2}}}\dfrac{\rho^2}{4}\,e^{-\rho/2}\sin\theta\cos\theta\,e^{i\phi}$
ψ_{320}	$\dfrac{1}{81\sqrt{(6\pi)}}\dfrac{1}{a^{\frac{3}{2}}}\dfrac{\rho^2}{4}\,e^{-\rho/2}(3\cos^2\theta - 1)$
ψ_{32-1}	$\dfrac{1}{81\sqrt{\pi}}\dfrac{1}{a^{\frac{3}{2}}}\dfrac{\rho^2}{4}\,e^{-\rho/2}\sin\theta\cos\theta\,e^{-i\phi}$
ψ_{32-2}	$\dfrac{1}{81\sqrt{(4\pi)}}\dfrac{1}{a^{\frac{3}{2}}}\dfrac{\rho^2}{4}\,e^{-\rho/2}\sin^2\theta\,e^{-i2\phi}$

not upon the individual values of its components l and s. Thus the levels $^2s_{\frac{1}{2}}$, $^2p_{\frac{1}{2}}$ both have $j = \frac{1}{2}$ and are energetically coincident. Similarly for $^2p_{\frac{3}{2}}$ and $^2d_{\frac{3}{2}}$. The transition probability between an initial stationary state 1 to a second 2 is given by

$$P_{12} = e\int r\psi_1^*\psi_2\,d\tau \qquad (4.7.17)$$

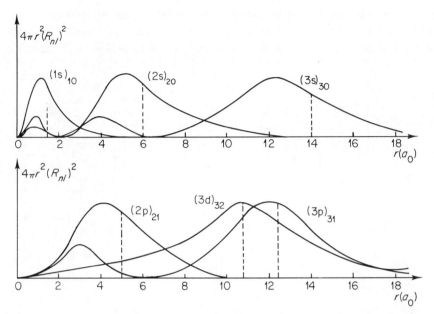

Figure 4.7.1 The normalized radial distributions of electronic charge in the hydrogen atom. The location of the corresponding Bohr orbits is shown by the broken lines

and if this is evaluated for the various states of the hydrogen atom it is found that the transition moment P_{12} is zero unless the two states concerned have orbital angular momentum quantum numbers l differing by 1. That is, the selection rule

$$\Delta l = \pm 1$$

operates. This is physically plausible since the condition of conservation of angular momentum confers unit angular momentum on the emitted photon, and this has been directly confirmed. Transitions Δn may, of course, adopt any positive or negative integral value, including zero. A comparison of the original Bohr and Dirac energy levels incorporating electron spin are shown in Figure 4.7.2. The allowed transitions with their associated intensities and frequencies are also indicated. The single Bohr level corresponding to the Balmer Hα line has five components on the Dirac theory. More recently, the interaction between an electron and a quantized electromagnetic field has been considered, and a removal of the Dirac degeneracy obtained. The field of quantum electro-dynamics is beyond the scope of the present book, but a qualitative explanation has been given in terms of the zero point energy associated with the quantized electromagnetic (i.e. Coulombic) field. The electron, subjected to these zero point effects, no longer remains a point charge, but delocalizes and may be considered as a *sphere* of charge which, in the hydrogen atom, means that it is not so strongly attracted to the nucleus. Consequently states of zero orbital angular momentum ($l = 0$) are *raised* in energy relative to those which are less likely to be found near the nucleus, and for which, therefore, the electron again approximates to a point charge with respect to the central field. These shifts, known as *Lamb shifts* are shown in Figure 4.7.2. The effect on states of higher

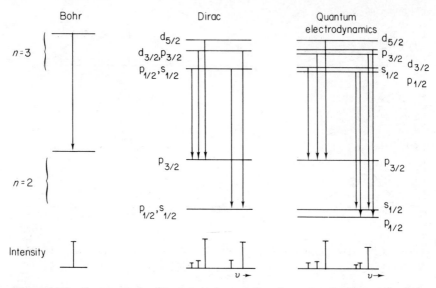

Figure 4.7.2 Energy levels of hydrogen for $n = 2$ and $n = 3$ according to the Bohr, Dirac and quantum electrodynamic theories

orbital angular momentum and principal quantum number is variable, either upwards or downwards depending on the time the delocalized electron spends in the vicinity of the nucleus. The Dirac degeneracy is eliminated and the multiplet, subject to the usual selection rules, now contains seven components. The multiplet structure of the 1640 Å line of He^+ obtained by Herzberg is shown in Figure 4.7.3.

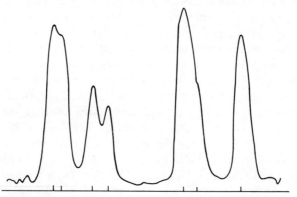

Figure 4.7.3 The multiplet structure of the 1640 Å line of He^+

4.8 Hydrogenic Systems

The theory of the hydrogen atom may be more or less directly applied to a number of systems including ionized atoms, mesic atoms, excitons and the alkalis. Clearly, in the case of singly ionized helium, doubly ionized lithium, etc., the electron–nuclear interaction is essentially a two-body system, and the resulting term diagram will be hydrogen-like. Since the energy of the nth hydrogenic eigenstate is (equation (4.7.11))

$$E_n = -\frac{Z^2 e^4 m_e}{2n^2 \hbar^2}$$

the energy levels for He$^+$ ($Z = 2$) will evidently be expanded by a factor of 4, whilst the coordinate scale of the wavefunctions will be reduced by a factor of 2. This hydrogenic series, first discovered in the star ζ-Puppis by Pickering, was initially attributed to some form of hydrogen. It was, however, subsequently shown that the spectrum arises from helium and is unobservable in pure hydrogen.

The phenomenon of photoconductivity in a semiconductor arises from the optical excitation of an electron from a full valence band to the conduction band, provided of course the energy of the photon is greater than the energy gap E_g between the valence and conduction bands. Both carriers, the electron and its associated hole, may move independently and participate in the conduction process. It may happen, however, that the photon energies are insufficient to excite the electron into the conduction band. In this case a neutral, mobile excited bound state may develop arising from the Coulombic interaction between the electron and hole. This *exciton* cannot participate in the conduction process, but can transport exciton energy within the crystal. These transient states represent a hydrogenic system of energy levels: this is confirmed by the selective optical absorption and emission spectrum (Figure 4.8.1). In this weakly bound or Mott–Wannier exciton the interparticle distance is large in comparison with atomic dimensions. (Another, tightly bound, exciton due to Frenkel, arises in molecular structures and molecular crystals. These exciton levels are much more closely related to the spectroscopic properties of the molecule than to the hydrogenic model under discussion here.) Since the masses of the electron and hole are identical we must use the reduced mass $\mu = m_e m_h/(m_e + m_h) = m_e/2$. Otherwise the energy levels are quite hydrogenic. A very similar transient system is *positronium* consisting of an electron–positron pair. For the brief period before the electron and positron annihilate in the form of γ-rays the stationary hydrogenic energy levels have been observed. Again, the reduced mass must be used, and as in the exciton case the energy scale shrinks by a factor of 2, whilst the wavefunctions are expanded by this factor.

Another transient system is the *mesic atom* formed from the negative μ-meson and the nucleus. Quite clearly the energy levels and wavefunctions will be hydrogenic for the brief period before it is captured by the nucleus, although the energy scale will be expanded by a factor of 212, whilst the coordinate scale will

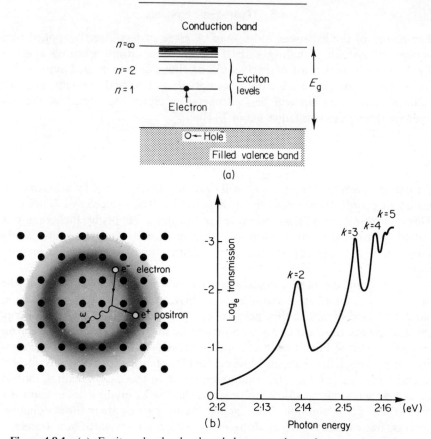

Figure 4.8.1 (a) Exciton levels developed between the valence and conduction bands. The electron and hole may recombine giving sharp optical transitions. (b) The spatially delocalized form of the weakly bound Mott–Wannier exciton extending over several lattice sites. The lifetime of the exciton is short, and the electron and hole recombine with the creation of an elementary excitation—a photon, phonon or magnon. The inverse process, the absorption of a photon creating an exciton, yields the absorption spectrum shown for Cu_2O at 77 K. (After P. W. Baumeister, *Phys. Rev.*, **121**, 359 (1961))

collapse by this factor. This is so since the mass of the μ-meson is $212\,m_e$. The μ-meson orbitals are therefore much closer into the nucleus than is the case for electrons, and the mesons consequently move in the field of an 'extended' source—the nucleus no longer being a simple point source of a Coulombic field. There is therefore a certain discrepancy between the mesonic and hydrogenic levels, particularly so for the heavier atoms where the meson orbitals are particularly close to the nucleus. These discrepancies have been observed in the mesic X-rays emitted from mesic transitions. The discrepancy enables deductions of the nuclear size and charge distribution to be made.

Perhaps the greatest deviation from the hydrogenic series develops in the alkalis having one valence electron. The single valence electron, however, penetrates the closed core states to a varying extent, and in consequence we are unable to speak of the electron as effectively moving in a screened field of a point charge $+e$. We may approximate the potential interaction of the single valence electron to the hydrogenic field by writing

$$V(r) = -\frac{e^2}{r}\left(1 + \frac{\beta}{r}\right) \tag{4.8.1}$$

where β is a constant. This model expresses the increase in effective nuclear charge as the electron penetrates the core. The radial equation using the above potential then becomes

$$\frac{d^2R}{dr^2} + \frac{2m_e}{\hbar^2}\left[E + \frac{e^2}{r} + \frac{e^2\beta}{r^2} - \frac{\hbar^2}{2m_e}\frac{l(l+1)}{r^2}\right]R = 0 \tag{4.8.2}$$

and defining

$$l'(l'+1) = l(l+1) - \frac{2m_e e^2 \beta}{\hbar^2}$$

we regain the usual radial equation

$$\frac{d^2R}{dr^2} + \frac{2m_e}{\hbar^2}\left[E + \frac{e^2}{r} - \frac{\hbar^2}{2m_e}\frac{l'(l'+1)}{r^2}\right]R = 0 \tag{4.8.3}$$

The energy eigenvalues correspondingly become

$$E_n = \frac{-Z^2 e^4 m_e}{2\hbar^2(n - \Delta_l)^2} \qquad \begin{matrix} n = 1, 2, 3 \\ l = 0, 1, \ldots, n - 1 \end{matrix}$$

where $\Delta_l = l - l'$ and is known as the *quantum defect*. We see immediately that the perturbation of the simple non-penetrating potential serves to remove the $(2l + 1)$-fold degeneracy—the energy eigenvalues now depending upon both n and l. Since the quantum defect is positive it serves to depress the energy levels below their hydrogenic values, the depression being the greatest for the orbits of low l and n, these penetrating the core states to the greatest extent. The absorption spectrum of ultraviolet light in lithium vapour is shown in Figure 4.8.2. Experimental values of the quantum defect for lithium determined from a spectroscopic analysis of the type shown in Figure 4.8.2 are given in Table 4.8.1.

Table 4.8.1. Effective quantum number n_{eff} and quantum defect Δ_l for lithium[a]

	$n = 2$	$n = 3$	$n = 4$	$n = 5$	$n = 6$	$n = 7$
n_{eff}	1·588	2·596	3·597	4·599	5·599	6·578
Δ_0	0·412	0·404	0·403	0·401	0·401	0·422

[a] The frequency of the absorption line is given as $v_n = T_\infty - R_{Li}/(n - \Delta_l)^2$ where T_∞ is the series limit and R_{Li} is the appropriate Rydberg constant. The results in the above table are obtained from the absorption spectrum in Figure 4.8.2 (a) taking $T_\infty = 43,487\ \text{cm}^{-1}$ and $R_{Li} = 109,729\ \text{cm}^{-1}$. The transitions involved in this table are all for the sharp series $T_\infty \to ns$.

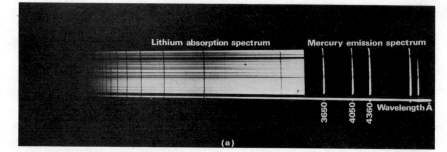

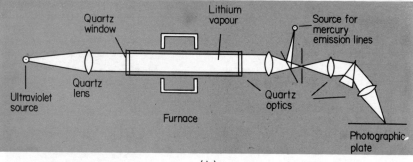

(b)

Figure 4.8.2 (a) The absorption spectrum of lithium. (b) The ultraviolet absorption spectrum of lithium vapour is recorded on the photographic plate on the right. The plate is calibrated by superimposing the known lines of the mercury spectrum: the frequency of the lines in the absorption series may then be simply determined by direct measurement of the photographic plate. The effective quantum numbers and quantum defects for the sharp series are given in Table 4.8.1

4.9 Spherical Harmonic Oscillator

A second example of a central force field for which there is an exact solution to the Schrödinger equation concerns the spherical harmonic oscillator which has the radial potential $V(r) = \frac{1}{2}m\omega^2 r^2$, where $\omega/2\pi$ is the classical frequency of oscillation. The radial equation correspondingly becomes

$$\frac{d^2R}{dr^2} + \frac{2}{r}\frac{dR}{dr} + \left(\frac{2mE}{\hbar^2} - \frac{m^2\omega^2 r^2}{\hbar^2} - \frac{l(l+1)}{r^2}\right)R = 0 \qquad (4.9.1)$$

Setting

$$\rho = m\omega r^2/\hbar, \quad \lambda = E/\hbar\omega, \quad R(r) = \mathscr{R}(\rho)$$

we obtain the somewhat simpler equation (cf. equation (4.6.5))

$$\frac{d^2\mathscr{R}}{d\rho^2} + \frac{1}{\rho}\frac{d\mathscr{R}}{d\rho} + \left(\frac{1}{4} - \frac{1}{2}\frac{\lambda}{\rho} - \frac{l(l+1)}{4\rho^2}\right)\mathscr{R} = 0 \qquad (4.9.2)$$

We again find that solutions finite at infinity are obtained only for

$$\lambda = 2k + l + \tfrac{3}{2}; \quad n = 2k + l + 1 = 1, 2, 3, \ldots$$
$$k = 1, 2, 3, \ldots$$

(4.9.3)

This establishes the quantization of the energy to the following discrete values

$$E_n = \hbar\omega(n + \tfrac{3}{2}) \tag{4.9.4}$$

The spacing of the energy levels is $\hbar\omega$ as in the case of the linear harmonic oscillator, except that now the ground state is $\tfrac{3}{2}\hbar\omega$ above the potential minimum instead of $\tfrac{1}{2}\hbar\omega$. For a two-dimensional oscillator the ground state would be $\hbar\omega$ above the minimum. This is of course a zero point effect, and a similar line of reasoning to that given in Section 2.8 establishes that this result is consistent with the uncertainty principle.

Again the degeneracy is extensive, there being $2l + 1$ different values of m, each having the same energy. Moreover, since the energy is dependent only upon the index n (equation (4.9.4)) it follows that the various allowed values of l (odd l if n is even, and vice versa) will all have the same energy. Thus, when n is odd there will be $\tfrac{1}{2}n - \tfrac{1}{2}$ different values of the index l, whilst if n is even there are $\tfrac{1}{2}n$. Consequently the order of the degeneracy is $\tfrac{1}{2}n(n + 1)$.

If we consider the large ρ form of equation (4.9.2) we obtain

$$\frac{d^2\mathscr{R}}{d\rho^2} + \frac{1}{4}\mathscr{R} \sim 0 \tag{4.9.5}$$

and the long range form of $\mathscr{R}$ is consequently of the form $B \exp(-\rho/2)$, if the solution is to be finite at infinity. The final radial solution is expressed in terms of the associated Laguerre functions of half-integral order:

$$R(r) = Cr^{l+1} \exp\left(-\frac{1}{2}\frac{m\omega r^2}{\hbar}\right) L_k^{l+\frac{1}{2}}\left(\frac{m\omega r^2}{\hbar}\right) \tag{4.9.6}$$

The angular solutions develop in terms of the associated spherical harmonics as usual, so that the final normalized eigenfunction for the harmonic potential is

$$\psi_{lmn}(r, \theta, \phi) = A(\beta r^2)^{\frac{1}{2}(l+1)} e^{-\frac{1}{2}\beta r^2} L_k^{l+\frac{1}{2}}(\beta r^2) P_l^m(\cos\theta) e^{\pm im\phi} \tag{4.9.7}$$

where A is a normalization constant, and $\beta = m\omega/\hbar$. As we might expect, application of a magnetic or electric field to the oscillator will remove a certain amount of degeneracy, particularly in the index m.

4.10 Quantization of the Motion of a Charged Particle in a Magnetic Field

In the solution of wave equation in general and the Schrödinger equation in particular, we have been able to establish the development of extensive degeneracy in the eigenstates, this being generally attributed to a high degree of symmetry in the system Hamiltonian. If the symmetry is destroyed then in many cases the degeneracy is completely, or at least partially, removed. Here we shall investigate the removal of the $(2l + 1)$-fold degeneracy in the eigenstate of a hydrogenic valence electron by the application of an external magnetic field. In the unperturbed system, of course, the valence electron executes an energetically quantized motion in the spherically symmetric central field of the nucleus. The extensive symmetry of the system results in the energy eigenvalues being independent of the quantum number m (Section 4.7):

$$E_{nlm} = -\frac{Z^2 e^4 m_e}{2\hbar^2 n^2}$$

where $n = l + 1, l + 2, \ldots$ is the principal quantum number. Since m may adopt integral values within the interval $-l \leqslant m \leqslant +l$, the degeneracy is clearly $(2l + 1)$-fold. Moreover we have not, as yet, accounted for the nuclear and electron spin: these will serve to increase the degeneracy of the eigenstate still further.

If now an external magnetic field is applied, the valence electron will execute a quantized motion in the combined central electrostatic and external magnetic fields. If we designate the unperturbed Hamiltonian operator $\mathscr{H}^{(0)}$, then in the presence of a magnetic field the motion is determined by the Hamiltonian

$$\mathscr{H} = \mathscr{H}^{(0)} + \lambda \mathscr{H}^{(1)} \tag{4.10.1}$$

where λ is an infinitesimal parameter and is separated from the finite part of $\mathscr{H}^{(1)}$ for the purpose of determining infinitesimals in power series expansions. The perturbed wavefunction $\psi_n^{(1)}$ may, for small perturbations, be expanded in terms of a complete set of orthonormal unperturbed functions in the usual way:

$$\psi_n^{(1)} = \psi_n + \sum_{n \neq i} a_{ni} \psi_i \tag{4.10.2}$$

where the coefficients $a_{ni} \ll 1$. Then, working only to first order in small quantities we have

$$E_{nlm} = \int \psi_{nlm}^* \mathscr{H}^{(0)} \psi_{nlm} \, d\tau + \lambda \int \psi_{nlm}^* \mathscr{H}^{(1)} \psi_{nlm} \, d\tau$$

$$= E_{nlm}^{(0)} + \lambda \int \psi_{nlm}^* \mathscr{H}^{(1)} \psi_{nlm} \, d\tau \tag{4.10.3}$$

where $\psi_{nlm} = R_{nl}(r) Y_l^m(\theta, \phi)$ is the unperturbed wavefunction.

If now we specifically consider the application of a weak magnetic field H_z in the positive z-direction, then from equation (4.10.1) we have

$$\mathscr{H}^{(1)} = M_z = (\hbar/i)(\partial/\partial\phi)$$

which is the z-component of the angular momentum operator, and $\lambda = -He/2m_e c$.

The expectation energy (4.10.3) then becomes

$$E_{nlm} = E_{nlm}^{(0)} - \frac{H_z e}{2m_e c} \int \psi_{nlm}^* M_z \psi_{nlm} \, d\tau \tag{4.10.4}$$

Now, since $Y_l^m(\theta, \phi)$ is an eigenfunction of the operator M_z with eigenvalue $m\hbar$, we have directly

$$E_{nlm} = E_{nlm}^{(0)} - \frac{meh H_z}{2m_e c} \qquad (4.10.5)$$

The energy now depends explicitly upon the eigenvalue m, and the $2(2l + 1)$-fold degeneracy of the energy level (n, l) (not forgetting the double degeneracy of the spin) in the absence of a field is thus removed. Equation (4.10.4) is a statement of the quantized expectation values of the z-component of the angular momentum vector. For a given value of l this amounts to the fact that the angular momentum vector associated with the orbital adopts only certain quantized projections onto the z-axis: in other words a *space quantization* develops in the presence of a magnetic field. In its absence the orientation of the vectors is isotropic and the degeneracy is regained.

This latter description, of course, formed the basis of the earlier *ad hoc* vector model of the atom which successfully explained the various Zeeman effects. In this model the magnetic dipole associated with the orbital motion of the electron precessed about the direction of the external magnetic field H_z such that the projection of the *total* angular momentum vector $\mathbf{j} = \mathbf{l} + \mathbf{s}$ onto the z-axis was quantized into $2(2l + 1)$ discrete values (Figure 4.10.1).

Both on the basis of the vector model and the subsequent quantum mechanical analysis given above, the multiplet levels are equally spaced at $ehH_z/2mc$, being integral multiples of the Bohr magneton $eh/2mc$ in the magnetic field. The energy levels then split as shown in Figure 4.10.2, the degeneracy having been removed and transitions between these multiplets may occur subject to certain selection rules.

In fact, as we have already mentioned, there is a two-fold increase in the degeneracy in the absence of a magnetic field by virtue of the two spin states of the valence electron, not to mention the further degeneracy associated with the nuclear spin. In the case of the electron spin there is the so-called spin–orbit coupling between $\mathbf{l}$ and $\mathbf{s}$, the orbit and spin vectors, to form a resultant total spin–orbit vector $\mathbf{j}$. It is this vector which precesses about the direction of the magnetic field in the vector model of the atom, although at sufficiently high field strengths $\mathbf{l}$ and $\mathbf{s}$ may decouple to precess independently about H_z, resulting in the *Paschen–Back* effect. For many electron systems the spin vectors may independently combine to form a total spin vector $\mathbf{S}$, and similarly for the orbit vectors $\mathbf{L}$, whereupon the Russell–Saunders or $\mathbf{LS}$ coupling mode governs the distribution of energy eigenvalues in the magnetic field. Alternatively the $\mathbf{j}$-vectors associated with each electron may develop independently, and then the individual $\mathbf{j}$-vectors will couple—the $\mathbf{j}$–$\mathbf{j}$ coupling scheme. In fact some intermediate scheme is generally observed. The spin vector associated with the nucleus may couple with the spin–orbit vectors, or, in the presence of a strong magnetic field will decouple and precess independently about H_z. This is the Back–Goudsmit effect. In each case, however, the extensive degeneracy of the

178

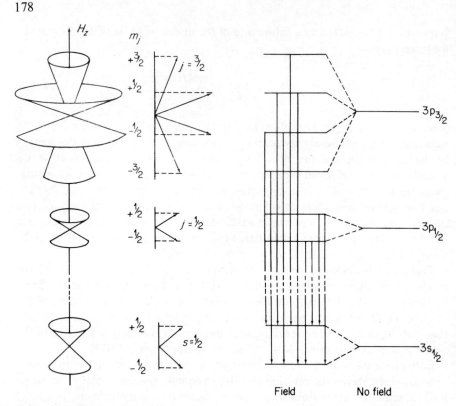

Figure 4.10.1 The precessional orientations of the orbital angular momentum vector of the 3p and 3s electrons of sodium in an external magnetic field

Figure 4.10.2 The Zeeman transitions for sodium. The transitions satisfy the selection rule $\Delta m = 0, \pm 1$

eigenstates is revealed, although whether the spectral splitting can always be resolved is another matter.

The analogous Stark effect, arising in the presence of electric fields, yields a similar resolution of the degeneracy.

4.11   The Liquid Metal Pseudopotential

An extremely interesting problem, related in many ways to the scattering and diffraction theory discussed in preceding sections, concerns the distribution of conduction electrons about the ionic core in a condensed metallic system. Once the electronic density distribution $\rho(r)$ is established it is relatively easy to relate this to the potential distribution via the Poisson equation: $\nabla^2 \psi(r) = -4\pi e^2 \rho(r)$.

It might be thought at first that the determination of the potential function would be impossibly complicated, depending upon all the core electrons as well as the conduction electrons. In fact, as Phillips and Kleinman observed in 1959, the process by which the conduction electrons are 'Pauli excluded' from the filled core states can be simulated mathematically by adding a repulsive component to the true ionic potential, Figure 4.11.3. The resulting effective potential —the pseudopotential—is rather weak, and whilst this enables it to be used in a whole range of perturbation calculations, it does, in particular, enable us to use the conventional theory of diffraction for the interaction of conduction electronic wavefunctions with an essentially 'opaque' ionic core. In a condensed assembly of ions we make the tacit assumption that the system is isotropic, and this is conveniently achieved if we restrict the discussion to a bulk liquid metal system.

Pronounced oscillations in the electronic density distribution about the core, and hence also in the potential, are obtained just as in the optical diffraction of monochromatic light about an opaque disc (Figure 4.11.1; cf. Figure 3.4.2).

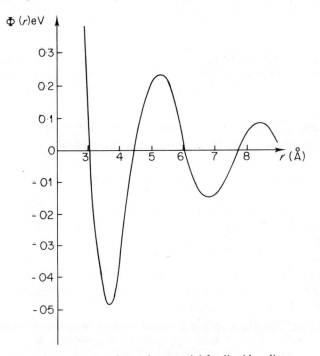

Figure 4.11.1 The ion–ion pair potential for liquid sodium showing the long range Friedel oscillations

This condition of monochromatism is fulfilled in the case of electron scattering in as far as the electrons participating in the conduction process are primarily those within kT of the Fermi surface. This implies reasonable 'monochromatism' of the electronic wavefunctions. We might anticipate that as the temperature is raised the spread in the range of wave vectors $\Delta k \sim \sqrt{(2mkT)}/h$ about the Fermi surface value k_F will lead to a blurring of the oscillatory features of the liquid metal potential (Figure 4.11.2).

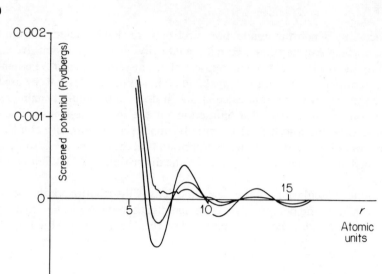

Figure 4.11.2 Variation of the pair potential for liquid mercury with increasing temperature. The curves show the effect of 0, 10 and 20% thermal blurring of the Fermi surface

We have seen that the exact solution to Schrödinger's equation for a spherical scattering potential may be expressed in terms of partial waves as follows:

$$\psi(r) \propto \sum_{l=0}^{\infty} j_l(kr)P_l^0(\cos\theta) \tag{4.11.1}$$

which for large r adopts the asymptotic form

$$\sum_{l=0}^{\infty} \frac{1}{kr}\sin(kr - \tfrac{1}{2}l\pi + \delta_l)P_l^0(\cos\theta) \tag{4.11.2}$$

We recall that the details of the scattering potential are incorporated in the phase shift δ_l: in the absence of a scattering potential, of course, $\delta_l = 0$ and equation (4.11.2) reverts to a partial wave description of a plane unscattered wave. If the scattering centre is the ionic core of radius r_0, then we require $\psi(r)$ to vanish at r_0. This immediately places the condition on k (from equation (4.11.2))

$$kr_0 + \delta_l = (n + \tfrac{1}{2}l)\pi \tag{4.11.3}$$

where n and l are integers.

The *change* in electron density at large r due to the presence of the scattering centre is, therefore, summing over all values of k up to k_F at the Fermi surface:

$$\Delta\rho(r) = \rho(r) - \rho_0$$
$$= \frac{1}{\Omega}\sum_{l=0}^{\infty}\int_0^{k_F}\sigma(k)\{|\psi(kr, \delta_l)|^2 - |\psi(kr)|^2\}k^2\,dk \tag{4.11.4}$$

where $|\psi(kr, \delta_l)|^2$, $|\psi(kr)|^2$ represent the probability distributions in the presence $(\delta_l > 0)$ and in the absence $(\delta_l = 0)$ of a scattering potential, respectively. $\sigma(k)$ is the density of states function which we have yet to specify, and Ω is the normalizing factor. This may, in fact, be shown to be $2\pi r_0$, at least in the absence of a scattering potential: we shall set $\Omega = 2\pi r_0$ throughout here.

From the condition on the allowed values of the wave vector (4.11.3), we see that for given l, a change $\Delta n = 1$ implies a change Δk given by

$$r_0 \Delta k + \Delta \delta_l = \pi$$

so that the density of states function is

$$\sigma(k) = \frac{1}{\Delta k} = \frac{2(2l + 1)}{\pi}\left(r_0 + \frac{\Delta \delta_l}{\Delta k}\right)$$

where, for a given energy or k-state there is a $(2l + 1)$-fold degeneracy associated with the orientation of the angular momentum, and a further 2-fold degeneracy associated with the electron spin. In fact $r_0 \gg d\delta_l/dk$ and we may write

$$\sigma(k) \sim \frac{2(2l + 1)}{\pi} r_0 \tag{4.11.5}$$

Thus, from equations (4.11.2), (4.11.4) and (4.11.5) we have

$$\Delta\rho(r) = \frac{1}{2\pi r_0}\sum_{l=0}^{\infty}\frac{2(2l + 1)}{\pi}r_0\int_0^{k_F}\frac{1}{(kr)^2}\{\sin^2(kr - \tfrac{1}{2}\eta l + \delta_l)$$

$$- \sin^2(kr - \tfrac{1}{2}l\pi)\}k^2\,dk \tag{4.11.6}$$

$$= \frac{1}{2\pi^2 r^3}\sum 2(2l + 1)\int_0^{k_F}\sin(2kr - l\pi + \delta_l)\sin\delta_l\,d(2kr)$$

$$= \frac{1}{2\pi^2}\sum(2l + 1)(-1)^{l+1}\sin\delta_l\cos\frac{(2k_F r + \delta_l)}{r^3}$$

In this treatment we have assumed a sharp cut-off at k_F, upon which the distribution depends explicitly as we anticipated. The assumption of a sharp Fermi surface applies, of course, only at absolute zero. Nonetheless, the essential feature of the electron distribution, and hence in the ion–ion potential, is the development of *Friedel oscillations* of wavelength $2k_F$, their amplitude falling off as r^{-3}. Thermal excitation and hence blurring of the Fermi surface at normally accessible temperatures is not expected to destroy the Friedel oscillations. Indeed, Gaskell and March have shown that a 20% blurring of the Fermi surface of Hg would be required to damp out the oscillatory effects of the interaction, and the blurring is usually much less than half this value. In Figure 4.11.2 we show some of the calculated curves of Gaskell and March.

So far we have considered the shielding of one ionic core in the Fermi sea of conduction electrons: our primary concern is, of course, with the pair interaction between ions. Now, the Coulomb interaction between two ions would be

182

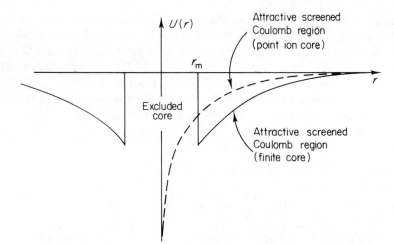

Figure 4.11.3 The 'empty core' model of the pseudo-potential for the electron–ion interaction. The potential simulates the effect of Pauli exclusion of the conduction electrons from the core states, and the screened Coulomb potential beyond the core radius r_m. The corresponding potential for the point-ion screened interaction is shown for comparison

$Z^2 e^2/r^2 \varepsilon$, where ε is the dielectric constant. It is possible to retain this classical formulation for the pair interaction and incorporate the wave theory result regarding the Friedel rather than Coulombic screening by allowing $\varepsilon \to \varepsilon(r)$. It is simpler from now on to work entirely in k-space so that the Fourier transform of the pair interaction becomes

$$U(k) = -\frac{N}{\Omega} \frac{4\pi Z^2 e^2}{k^2 \varepsilon(k)} \tag{4.11.7}$$

where by Fourier transforming our initial expression for the potential distribution (from Poisson's equation applied to equation (4.11.6)) we have

$$\varepsilon(k) = \frac{1}{k^2} \left[k^2 + \frac{k_F}{\pi a_0} g\left(\frac{k}{2k_F}\right) \right] \tag{4.11.8}$$

where

$$g(x) = 2 + \frac{x^2 - 1}{x} \ln \left| \frac{1 - x}{1 + x} \right|, \quad a_0 = \hbar^2/me^2$$

$$(NZ/\Omega) = k_F^3/3\pi$$

The schematic form of $U(k)$ is shown in Figure 4.11.4. We see from the equations (4.11.7), (4.11.8) that $U(k) \to -k_F^2 e^2 a_0/3 = -2E_F/3$ as $k \to 0$, and this is a central result of the theory. Another important feature is that $U(k)$ passes through zero as k approaches $2k_F$. The location of the zero, and the detailed form of $U(k)$ are, however, still under discussion.

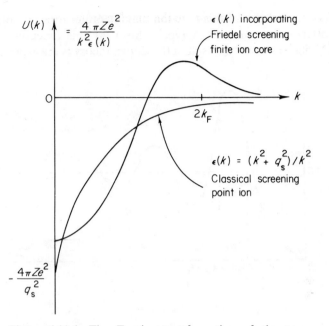

Figure 4.11.4 The Fourier transformation of the two pseudo-potentials shown in Figure 4.11.3. The important feature of the Friedel-screened finite core function $U(k)$ is the node at $k < 2k_F$ and the positive region at larger k: the precise location of this node, however, is the subject of some controversy

Ziman has applied the pseudopotential concept to the calculation of resistivities and thermoelectric powers of liquid metals. As we might expect, both the structure and the nature of the liquid metal scattering potential will affect the resistivity. In a perfectly ordered lattice at absolute zero, but for some disorder and impurity scattering, we would expect the metal to exhibit a negligible electrical resistance. With increasing temperature the lattice vibrations will scatter the conduction electrons and the resistance will rise. In this way we are able to explain the temperature dependence of the resistivity of pure metals and alloys. With the sudden decrease in order on melting it is somewhat surprising to find that the resistivity is relatively unaffected by the phase change. Related to this is the apparent success of the 'nearly free electron' model of electronic conduction according to which the array of ions in the metal only slightly influences the conduction properties. This latter effect we may understand when we recall the relatively weakly scattering pseudopotentials: the conduction electron wavefunctions are orthogonalized to the core states and the ion–electron interaction is correspondingly weak. The *arrangement* of the ions must, however, enter the discussion. Closely related to the Fourier transform of the radial distribution of ions or scattering centres is the *structure*

184

factor, $S(k)$. Knowledge of this function may be rather directly obtained from X-ray scattering experiments. A typical form for this function is shown in Figure 4.11.5(a). The central result of the Ziman theory is the expression for the

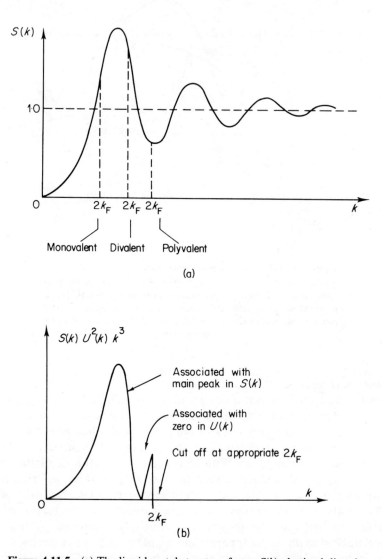

(a)

(b)

Figure 4.11.5 (a) The liquid metal structure factor $S(k)$ obtained directly from X-ray scattering experiments. The general form of this function does not vary greatly from system to system, although the oscillations broaden and dampen with increasing temperature. The location of the quantity $2k_F$ for mono-, di- and polyvalent metals is indicated. (b) The integrand $S(k)U^2(k)k^3$ arising in the resistivity integral. As we see, the area beneath this curve is sensitively dependent upon the location of the node in $U(k)$ and the cut-off $2k_F$

resistivity:

$$R = \alpha \int_0^{2k_F} S(k)|U(k)|^2 k^3 \, dk$$

The form of the integrand is shown in Figure 4.11.5(b) the area beneath the curve is proportional to the resistivity of the liquid. The location of the cut-off at $2k_F$ depends whether the liquid metal is mono-, di-or polyvalent. The structure factor $S(k)$ is, however, relatively independent of the liquid system. Since $2k_F(\text{poly}) > 2k_F(\text{di}) > 2k_F(\text{mono})$, generally, we may immediately explain for example the fact that polyvalent liquid metals (Pb, Sn) have a higher resistance than monovalent systems (Na, Ag). In fact, the cut-off value of $2k_F$ in silver occurs *before* the main peak in the structure factor. In zinc the cut-off happens to be located centrally on the first peak of $S(k)$. With increasing temperature the principal peak in $S(k)$ becomes lower and broader, and actually accounts for the *decrease* in the resistivity of zinc with increasing temperature.

Quite clearly the pseudopotential concept enables us to account at once for the nearly free electron behaviour in the liquid metal, and justifies the use of elementary diffraction theory in the discussion of the electrical properties.

4.12 Propagation of Earth Tremors at the Mohorovičic Layer

One of the current geophysical models of the earth considers the interior structure to consist of a central relatively mobile core of high density and low rigidity modulus, and an outer mantle of slightly lower density, but much higher rigidity modulus (Figure 4.12.1). It is, of course, the constant adjustment of this mantle, principally along fault lines and rifts, which is responsible for the earthquakes and tremors detected by seismological stations around the world. On the basis of the nature of the waves propagated through the earth the above model has been developed. The boundary between the two regions of differing rigidity modulus is termed the Mohorovičic layer and here we shall attempt to determine some of the characteristics of the normal modes of vibration of such a system.

We assume the radial variation of the rigidity modulus to vary in some arbitrary fashion $\mu(r) = \mu_0[1 - f(r)]$, whilst the density will be taken to be constant. The analysis is a little more lengthy, but no more difficult in principle if we included a radial density dependence, but here we shall assume it is constant. The velocity of propagation of compressional acoustic waves in the system is given as $\sqrt{(\mu/\rho)} = \sqrt{\{\mu_0[1 - f(r)]/\rho\}}$ whilst the modified eigenvalue $k_m^2 = k_m^{0\,2}(1 + \alpha_m)$, in terms of the unperturbed value designated by ^{0}. Both $f(r)$ and α_m are taken to be small quantities $\ll 1$ which restricts the application of this perturbative approach.

The nth order radial wave equation for the mth perturbed mode is written (cf. equation (4.5.11))

$$\frac{d^2 R_n}{dr^2} + \frac{1}{r}\frac{dR_n}{dr} + \left\{ \frac{4\pi v_m^2 \rho}{\mu_0[1 - f(r)]} - \frac{(n + \frac{1}{2})}{r^2} \right\} R_n = 0 \qquad (4.12.1)$$

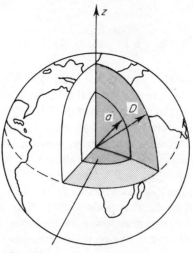

Core of
low rigidity
modulus

Figure 4.12.1 The discontinuity in
rigidity modulus between the outer
mantle and the core of the earth.
Seismic disturbances tend to propagate
around the outer mantle of higher
rigidity modulus than the core

i.e.

$$\frac{d^2 R_n}{dr^2} + \frac{1}{r}\frac{dR_n}{dr} + \left\{ k_m^{0\,2}[1 + \alpha_m + f(r)] - \frac{(n + \frac{1}{2})}{r^2} \right\} R_n = 0$$

where   (4.12.2)

$$k_m = 2\pi v_m \sqrt{(\rho/\mu)}$$

working only to first order in small quantities. If we now make a Fourier
expansion of the perturbed mode in terms of a complete set of orthogonal
spherical Bessel functions of the first kind, then

$$R_{n,m}(r) = A_{n,m} j_n(k_m^0 r) + \sum_{i \neq m}^{\infty} A_{n,i} j_n(k_i^0 r)$$ (4.12.3)

where the k_i^0 are chosen such that $(dj_n(k_i^0 D)/dr) = 0$: D = radius of earth. We
shall assume that the mth perturbed mode will most closely resemble the mth
*un*perturbed mode, but this is nevertheless hybridized with all other members
of the set. We assume therefore that $A_m \gg A_i$. Since we require finite interior
solutions, there is no question of the development of radial solutions of the
second kind.

Substituting equation (4.12.3) into equation (4.12.2), working only to first order in small quantities, and neglecting the product of small quantities, we obtain after some tedious algebraic manipulation

$$[\alpha_m + f(r)]k_m^{0\,2}A_{n,m}j_n(k_m^0 r) + \sum_{i \ne m}^{\infty} A_{n,i}(k_m^{0\,2} - k_i^{0\,2})j_n(k_i r) \qquad (4.12.4)$$

If we now multiply throughout by $rj_n(k_m^0 r)$, (where the weighting factor has been included; see Section 1.7), and integrate, we obtain

$$\alpha_m = -\frac{\int_0^D f(r) j_n^2(k_m^0 r)r \, dr}{\int_0^D j_n(k_m^0 r)r \, dr} \qquad (4.12.5)$$

This follows from the orthogonality property of the eigenfunctions, with respect to the weighting function, over the interval $0 \le r \le D$. Similarly, multiplying throughout by $rj_n(k_i^0 r)$ and integrating yields

$$\frac{A_{ni}}{A_{nm}} = \frac{k_m^{0\,2}}{k_m^{0\,2} - k_i^{0\,2}} \frac{\int_0^D f(r) j_n(k_m^0 r) j_n(k_i^0 r)r \, dr}{\int_0^D j_n^2(k_i^0 r)r \, dr} \qquad (4.12.6)$$

for the hybridization ratio. This gives the amplitude of the hybridization of the mth eigenfunction with the ith, relative to the amplitude of the mth harmonic. We notice that, as usual, hybridization takes place most extensively between adjacent modes having similar eigenvalues, that is those for which $k_m^0 \sim k_i^0$.

We have yet to specify the radial perturbation of the rigidity modulus, $f(r)$. This may be conveniently done by assuming that $f(r)$ is a step function (Figure 4.12.2a) having its discontinuity located on the Mohorovičic boundary at $r = a$. The corresponding variation of the rigidity modulus $\mu(r)$ is shown in Figure 4.12.2(b).

Expressions (4.12.5), (4.12.6) now become

$$\alpha_m = \frac{\int_a^D j_n^2(k_m^0 r)r \, dr}{\int_0^D j_n^2(k_m^0 r)r \, dr} \qquad (4.12.7)$$

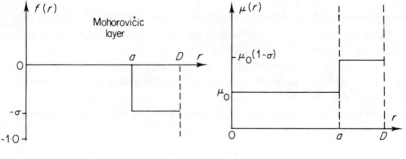

(a)   (b)

Figure 4.12.2

and

$$\frac{A_{ni}}{A_{nm}} = \frac{-k_m^{0\,2}\sigma}{k_m^{0\,2} - k_i^{0\,2}} \frac{\int_a^D j_n(k_m^0 r) j_n(k_i^0 r) r \, dr}{\int_0^D j_n^2(k_i^0 r) r \, dr} \tag{4.12.8}$$

We see that we recover the unperturbed solutions as $\sigma \to 0$.

Qualitatively we may reason that only those radial modes will be sustained which satisfy the more stringent requirement that the spherical Bessel function should have roots at a and D. The rigid mantle therefore acts as a high-pass filter disallowing eigenvalues below $k_{\min} \sim \pi/(D - a)$. The normal modes are, however, perturbed, and we see that a more correct expression of the suppression of low frequency modes comes from equation (4.12.8) which is seen to be zero whenever the functions $j_n(k_m r), j_n(k_i r)$ are orthogonal on the interval $a \leqslant r \leqslant D$. Just from a consideration of the limits we see from equation (4.12.7) that all frequencies are raised by the factor $(1 + \sigma(1 - a/D))^{\frac{1}{2}}$, very approximately. In fact the frequencies will be slightly lower than this as a closer examination of equation (4.12.7) will suggest.

We have computed the first few hybridization ratios relative to the fundamental A_1 from equation (4.12.8), and these are listed in Table 4.12.1 for various

Table 4.12.1. Hybridization coefficients with respect to the fundamental (A_1/A_m) for the spherical Bessel function $j_0(k_m r)$

m	$j_0(r)$ A_1/A_m	$j_1(r)$ A_1/A_m	$j_2(r)$ A_1/A_m
1	1·000000	1·000000	1·000000
2	−0·035396	−0·090255	−0·156605
3	0·018484	0·040039	0·060084
4	−0·008916	−0·015807	−0·018539
5	0·002613	0·002090	−0·001906
6	0·000954	0·004268	0·009263

harmonics k_m, these for the spherical Bessel functions, j_0, j_1, j_2. The resulting perturbations of the radial functions are shown in Figure 4.12.3 where we see that the amplitude is greatly enhanced beyond the Mohorovičic layer, whilst it is effectively suppressed in the core.

If a seismic disturbance occurs at a point on the earth's surface (D, θ, ϕ) at time $t = 0$, then setting the z-axis to pass through this point we may restrict the angular distribution of waves at any instant to the zonal set $Y_n^0(\theta)$. Then, if the disturbance is represented by a delta function $\delta(r - D, \theta)$ at the pole of coordinates we have, at $t = 0$,

$$\delta(r - D, \theta) = \sum_n^\infty \left\{ \sum_m^\infty A_{nm} j_n(k_m^0 r) + \sum_{i \neq m}^\infty A_{ni} j_n(k_i^0 r) \right\} P_n^0(\cos \theta) \tag{4.12.9}$$

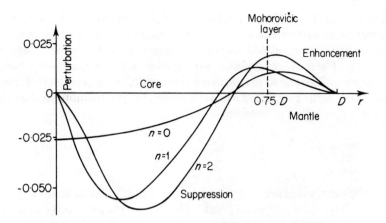

Figure 4.12.3 The difference $(j_n(\text{perturbed}) - j_n)$ represents the enhancement, or suppression of the nth spherical Bessel function due to the presence of the discontinuity in rigidity modulus. As we see in the diagram above, at radial distances beyond the discontinuity located at $0 \cdot 75D$ the oscillations are enhanced, whilst within the core they are suppressed. These calculations are for the fundamental modes of the functions j_0, j_1, j_2. $\sigma = -0.90$. The hybrid ratios are given in Table 4.12.1

We may determine the amplitudes A_{nm}, A_{ni} from the orthogonality properties of the j_n and P_n functions. Thus, multiplying both sides of equation (4.12.9) by $r j_n(k_m^0 r) P_n^0(\cos \theta)$ and integrating, we have

$$\int_0^D \int_0^\pi \delta(r - D, \theta) j_n(k_m^0 r) P_n^0(\cos \theta) r \sin \theta \, d\theta \, dr$$

$$= \int_0^D \int_0^\pi \sum_n \left\{ \sum_m A_{nm} j_n(k_m^0 r) + \sum_{i \neq m} A_{ni} j_n(k_i^0 r) \right\} \qquad (4.12.10)$$

$$\times P_n^0(\cos \theta) P_n^0(\cos \theta) j_n(k_m^0 r) r \sin \theta \, d\theta \, dr$$

i.e.

$$j_n(k_m^0 D) = A_{nm} \frac{2}{2n + 1} \int_0^D j_n^2(k_m^0 r) r \, dr$$

since

$$P_n^0(\cos 0) = 1 \quad (\theta = 0), \quad \int_0^\pi P_n(\cos \theta) P_m(\cos \theta) \sin \theta \, d\theta = \frac{2\delta_{mn}}{2n + 1}$$

where δ_{mn} is the Kronecker delta. So that

$$A_{nm} = \frac{(2n + 1)}{2D} j_n(k_m D) \left\{ \int_0^D j_n^2(k_m^0 r) r \, dr \right\}^{-1} \qquad (4.12.11)$$

A similar analysis yields the amplitudes A_{ni}. The eigenvalues $k_m = k_m^0(1 + \alpha_m)^{\frac{1}{2}}$ are determined as the roots of the Neumann boundary condition $(dj_n(z)/dz)_D = 0$, which is more complicated than the simple Dirichlet condition used at the boundary of a rotating gaseous star (Section 4.5). Physically the Neumann condition corresponds to the normal mode condition for reflection of *transverse* waves at the earth's surface. The final perturbed eigenfunctions for the spherical interior are therefore of the form

$$\psi(r, \theta, t) = \sum_{n=0}^{\infty} \left\{ \sum_{m=0}^{\infty} A_{nm} j_n(k_m^0 r) + \sum_{i \neq m}^{\infty} A_{ni} j_n(k_i^0 r) \right\} P_n^0(\cos \theta)\, e^{i k_m^0 c t}$$

$$(4.12.12)$$

where the amplitudes are given in equation (4.12.11). We see that the perturbed eigenvalues have been expressed entirely in terms of the unperturbed quantities ^{0}. There is some ambiguity regarding our choice of c, the velocity of propagation of elastic waves in the system. As it is, there will be two distinct velocities of propagation c_l and c_t corresponding to longitudinal and transverse waves. The longitudinal component requires the simpler Dirichlet boundary condition at $r = D$.

The condition at the free surface of an elastic solid body that there should be an absence of normal and tangential stress makes it clear that longitudinal or transverse waves cannot exist alone. Indeed, the reflection of a transverse wave from the boundary gives rise to a longitudinal and a transverse wave, similarly for the reflection of a longitudinal wave. The two reflected waves will generally have the same frequency, but different wavelengths in accordance with their velocities of propagation. It is clear that at the boundary some combination or superposition of longitudinal and transverse waves will develop, this combination satisfying the condition of vanishing normal and tangential stress at the free surface. The waves so developed may be dealt with as possessing a progressive component which spreads over the surface of the sphere, and a normal (radial) standing component. In addition to these, Rayleigh and Love waves may also develop at the free surface (see Section 3.6).

4.13 Interior Temperature Distribution for a Rotating Planet

An important problem concerning diffusive and thermal fields in spherical coordinates is the determination of the temperature distribution within a sphere due to external heating. The specific problem we shall consider is the interior temperature distribution for a planet rotating in the radiation field of the sun.

First, however, we shall consider a very simple radial diffusion problem. Given that the initial field distribution is $\psi(r, 0)$, the problem is to determine the subsequent evolution within the sphere, subject to the boundary condition

$$\psi(a, t) = 0 \qquad (4.13.1)$$

the sphere being of radius a.

As usual we assume a separation of variables $\psi(r, t) = R(r)T(t)$ and obtain the separated equations

$$\frac{dT}{dt} + k^2 DT = 0 \tag{4.13.2}$$

and

$$\frac{d^2R}{dr^2} + \frac{2}{r}\frac{dR}{dr} + k^2 R = 0 \tag{4.13.3}$$

where k is real.

Solutions to equation (4.13.2) are of the form

$$T(t) = Ce^{-k^2 Dt}$$

whilst if we make the substitution

$$R(r) = r^{-\frac{1}{2}}\mathscr{R}(r) \tag{4.13.4}$$

then equation (4.13.3) becomes

$$\frac{d^2\mathscr{R}}{dr^2} + \frac{1}{r}\frac{d\mathscr{R}}{dr} + \left(k - \frac{1}{4r^2}\right)\mathscr{R} = 0 \tag{4.13.5}$$

i.e.

$$R(r) = Ar^{-\frac{1}{2}}J_{\frac{1}{2}}(kr) = A\sqrt{\left(\frac{2k}{\pi}\right)}j_0(kr) = A'\frac{\sin kr}{r} \tag{4.13.6}$$

rejecting interior radial solutions of the second kind. The general solution to the diffusion equation is therefore

$$\psi(r, t) = \sum_{n=1}^{\infty} \frac{A_n}{r}\sin(k_n r)\exp(-k_n^2 Dt) \tag{4.13.7}$$

The boundary condition (4.13.1) enables us to set $k_n = n\pi/a$ where n is an integer (exploiting the harmonicity of the zero-order spherical Bessel function). Further, at $t = 0$ we have

$$r\psi(r, 0) = \sum_{n=1}^{\infty} A_n \sin\left(\frac{n\pi r}{a}\right) \tag{4.13.8}$$

and using the orthogonality of $\sin(n\pi r/a)$ on the interval $2a$ the coefficients A_n are determined as

$$A_n = -\frac{2}{a}\int_0^a r\psi(r, 0)\sin\left(\frac{n\pi r}{a}\right)dr \tag{4.13.9}$$

and

$$\psi(r, t) = \sum_{n=1}^{\infty} \frac{A_n}{r}\sin\left(\frac{n\pi r}{a}\right)\exp\left(-\frac{n^2\pi^2}{a^2}Dt\right) \tag{4.13.10}$$

The boundary requirement that $\psi(a, t) = 0$ serves only as a reference level with respect to the internal evolution. If the sphere is immersed in a constant ambient field ψ_0 it is only necessary to add this constant on to equation (4.13.10) to obtain the absolute rather than the relative field distribution.

We need not have solved the problem in quite this way. We could have left the solution in terms of the spherical Bessel function and determined the allowed values of k_n from the coincidence of the nodes of $j_0(kr)$ with the surface of the sphere. The arbitrary coefficients A_n could have been determined utilizing the orthogonality properties of the Bessel functions rather than of the sine functions, as in equation (4.13.9).

We now move on to consider the interior temperature distribution within a rotating planet immersed in the radiation field of the sun. The boundary condition is given by the rate of energy intake at the planet's surface. If we take the axis of rotation as the z-axis ($z = r \cos \theta$), and the sun to be located sufficiently far along the positive x-axis ($x = r \sin \theta \cos \phi$) for the radiation wavefront at the planet to be effectively planar, then the rate of energy intake over the planet's surface is $I_0 \sin \theta \cos \phi$ (Figure 4.13.1). If, however, we take axes *fixed with*

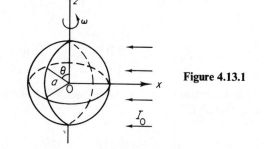

Figure 4.13.1

respect to the sphere then the inflow becomes $I_0 \sin \theta \cos (\omega t - \phi)$. Assuming a solution of the form $\psi = R(r) Y(\theta, \phi) e^{-i\omega t}$ the diffusion equation becomes

$$\nabla^2 \psi = -\frac{i\omega}{D} \psi \qquad (4.13.11)$$

and after the usual separation of variables we obtain the radial equation

$$\frac{d^2 R}{dr^2} + \frac{2}{r} \frac{dR}{dr} + \left(\frac{i\omega}{D} - \frac{l(l+1)}{r^2} \right) R = 0 \qquad (4.13.12)$$

which, after the transformation $R(r) = r^{-\frac{1}{2}} \mathcal{R}(r)$ becomes

$$\frac{d^2 \mathcal{R}}{dr^2} + \frac{1}{r} \frac{d\mathcal{R}}{dr} + \left(\frac{i\omega}{D} - \frac{(l(l+1) + \frac{1}{4})}{r^2} \right) \mathcal{R} = 0 \qquad (4.13.13)$$

From the symmetry of the problem it is evident that $l = 1$, whereupon we have the radial solution

$$\mathscr{R}(r) = AJ_{\frac{3}{2}}\left(\sqrt{\left[\frac{i\omega}{D}\right]}r\right) = \frac{A}{r^{\frac{1}{2}}}j_1\left(\sqrt{\left[\frac{i\omega}{D}\right]}r\right)$$

so that

$$R(r) = Aj_1\left(\sqrt{\left[\frac{i\omega}{D}\right]}r\right) \tag{4.13.14}$$

where j_1 is the first-order spherical Bessel function of the first kind. D has the significance of a thermal diffusivity, and $D = K/\rho C$ where k is the thermal conductivity, ρ the density and C the specific heat of the medium.

The distribution of temperature inside the planet is therefore that solution of equation (4.13.11) which has $k(\partial\psi/\partial r)_a$ equal to the specified energy influx. That is,

$$\psi = Aj_1\left(\sqrt{\left[\frac{i\omega}{D}\right]}r\right)e^{\pm i\phi}\sin\theta\, e^{\pm i\omega t}$$

$$k\left(\frac{\partial\phi}{\partial r}\right)_a = kA\sqrt{\left[\frac{i\omega}{D}\right]}j_1'\left(\sqrt{\left[\frac{i\omega}{D}\right]}a\right)e^{\pm i\phi}e^{\pm i\omega t}\sin\theta = I_0\sin\theta\cos(\omega t - \phi)$$

$$\tag{4.13.15}$$

where j_1' is the first derivative of j_1 with respect to its argument. Equation (4.13.15) enables us to determine the coefficient A, and write the final solution as

$$\psi(r, \omega) = \operatorname{Re}\frac{I_0 j_1(\sqrt{[i\omega/D]}r)\sin\theta}{k\sqrt{[i\omega/D]}j_1'(\sqrt{[i\omega/D]}a)}\cos(\omega t - \phi) \tag{4.13.16}$$

where Re signifies that we are interested in the real part only. It should be pointed out that this solution assumes that there is no net inflow or outflow of heat: energy which is absorbed on one side is re-radiated on the other. Any source or sink terms would require the inclusion of a non-periodic component in the initial diffusion equation. Equation (4.13.16) nevertheless probably represents quite an adequate approximation for a planet in equilibrium with the radiative field.

If the planet rotates slowly such that the argument $a\sqrt{(\omega/D)}$ is small, then $j_1(x) \sim \frac{1}{3}x$ and $j_1'(x) \sim \frac{1}{3}$ (see Appendix III.16). Equation (4.13.16) then reduces to

$$\psi(r, \omega) \sim \frac{I_0}{k}r\sin\theta\cos(\phi - \omega t) \tag{4.13.17}$$

and the temperature gradient is therefore uniform across the sphere. In the limit as $\omega \to 0$ so the planet will adopt a uniform temperature distribution in equilibrium with the sun's radiation field.

In the converse problem of a rapidly rotating planet such that the argument $a\sqrt{(\omega/D)}$ is large, the spherical Bessel function $j_1(x) \to -(1/x)\cos x \sim -(1/2x)\,e^{ix}$ for r near the surface $(r \sim a)$ (see Appendix III.17). Further, $j_1(x) \sim (i/2x)\,e^{-ix}$. Equation (4.13.16) in this approximation becomes

$$\psi(r, \omega) = \mathrm{Re} \frac{I_0}{ik\sqrt{[i\omega/D]}} \frac{a}{r} \frac{\exp - (i\sqrt{[i\omega/D]}r)}{\exp - (i\sqrt{[i\omega/D]}a)} \sin\theta \, e^{i(\phi - \omega t)} \tag{4.13.18}$$

$$= \mathrm{Re} \frac{-I_0}{ik\sqrt{[i\omega/D]}} \exp - \left\{ i\sqrt{\left[\frac{i\omega}{D}\right]}(r - a) \right\} \exp i(\phi - \omega t) \sin\theta \tag{4.13.19}$$

since $r/a \sim 1$, and $\sqrt{i} = 1/\sqrt{2} + i/\sqrt{2}$, the temperature distribution is

$$= \frac{I_0}{k\sqrt{[i\omega/D]}} \exp - \left\{ \sqrt{\left[\frac{\omega}{2D}\right]}(r - a) \right\} \cos\left(\omega t - \sqrt{\left[\frac{\omega}{2D}\right]}(r - a) - \phi - \frac{\pi}{4} \right) \sin\theta \tag{4.13.20}$$

Notice that the thermal wavefront rotates with ϕ as the planet rotates, but $\pi/4$ out of phase. The wavelength of the distribution is $\sqrt{(8\pi^2 D/\omega)}$ and the wave velocity is $\sqrt{(2D\omega)}$. The waves attenuate rapidly beneath the surface of the planet, reducing by a factor of $e^{-2\pi}$ with each wavelength. The two special cases arise as limiting conditions on the relative magnitudes of the two characteristic periods corresponding to rotation and thermal diffusion—in the first case $\tau_{rot} \ll \tau_{diff}$ and in the second $\tau_{rot} \gg \tau_{diff}$.

4.14 Prolate and Oblate Spheroidal Coordinates

We shall very briefly consider the spheroidal coordinates and introduce the mathematical functions that generally arise in the solution of the Laplace and Helmholtz equations in this coordinate frame. Since the prolate and oblate coordinate systems arise from the rotation of the elliptic coordinates about the major and minor axis respectively, a sound understanding of the elliptic solutions is an essential prerequisite.

For convenience in the discussion of the prolate system we take the foci to be located at $z = \pm a/2$, thereby establishing the axial rotation coordinate ϕ about the z-axis as usual. The location of a point in the frame is given by the coordinate $(\cosh\mu, \cos\theta, \phi)$ which differs from the elliptic case only in the rotational coordinate ϕ (cf. Figure 4.14.1). Generally the coordinates are represented in the abbreviated nomenclature as follows:

$$\left. \begin{array}{l} \phi = \phi \\ \eta = \cos\theta = (r_1 - r_2)/a \\ \xi = \cosh\mu = (r_1 + r_2)/a \end{array} \right\} \tag{4.14.1a}$$

The distances r_1, r_2 and a are shown in Figure 4.14.1.

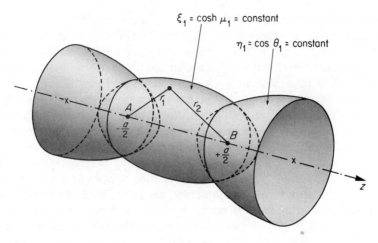

Figure 4.14.1

The replacement in terms of η, ξ achieves some simplification in the resulting expressions for the field operators ∇, ∇^2, etc. In the elliptic case the corresponding expressions were sufficiently simple that we were able to retain the $\cos \theta$, $\cosh \mu$ nomenclature throughout. We should notice that the range of the angular variable is $+1 \leqslant \eta \leqslant -1$, i.e. $2\pi \leqslant \theta \leqslant 0$, whilst that of the radial variable is $+\infty \leqslant \xi \leqslant +1$, i.e. $+\infty \leqslant \mu \leqslant 0$. The surface $\xi = $ constant is a prolate spheroid of interfocal distance a, whilst for $\eta = $ constant we obtain orthogonal *hyperboloids of two sheets* again having an interfocal distance of a, shown in Figure 4.14.1. In these coordinates we have

$$\nabla = \hat{\xi}\frac{2}{a}\sqrt{\frac{\xi^2 - 1}{\xi^2 - \eta^2}}\frac{\partial}{\partial\xi} + \hat{\eta}\frac{2}{a}\sqrt{\frac{1 - \eta^2}{\xi^2 - \eta^2}}\frac{\partial}{\partial\eta} + \frac{\hat{\phi}}{a}\frac{2}{\sqrt{[(\xi^2 - 1)(1 - \eta^2)]}}\frac{\partial}{\partial\phi}$$

$$(4.14.1)$$

and

$$\nabla^2 = \frac{4}{a^2(\xi^2 - \eta^2)}\left\{\frac{\partial}{\partial\xi}(\xi^2 - 1)\frac{\partial}{\partial\xi} + \frac{\partial}{\partial\eta}(1 - \eta^2)\frac{\partial}{\partial\eta} + \frac{\xi^2 - \eta^2}{(\xi^2 - 1)(1 - \eta^2)}\frac{\partial^2}{\partial\phi^2}\right\}$$

$$(4.14.2)$$

where $\hat{\xi}$, $\hat{\eta}$ and $\hat{\phi}$ are the unit vectors. If we assume a separable solution of the form $\psi(\xi, \eta, \phi) = R(\xi)\Theta(\eta)\Phi(\phi)$, then Laplace's equation separates to give

$$\frac{d}{d\xi}\left\{(1 - \xi^2)\frac{dR}{d\xi}\right\} + \left(l(l + 1) - \frac{m^2}{1 - \xi^2}\right)R = 0 \qquad (4.14.3)$$

$$\frac{d}{d\eta}\left\{(1 - \eta^2)\frac{d\Theta}{d\eta}\right\} + \left(l(l + 1) - \frac{m^2}{1 - \eta^2}\right)\Theta = 0 \qquad (4.14.4)$$

$$\frac{d^2\Phi}{d\phi^2} + m^2\Phi = 0 \qquad (4.14.5)$$

Dealing with these equations in reverse order, equation (4.14.5) will, of course, have the usual trigonometric solutions for real m, in which case rotational continuity requires that m be an integer. Recalling that the variable $\eta = \cos \theta$ has a range between $+1$ and -1, finite solutions to equation (4.14.4) are obtained only if l is zero or a positive integer. It follows that equation (4.14.4) is the associated Legendre equation and that it has solutions of the first kind, $P_l^m(\cos \theta)$. Equation (4.14.3) is not quite so straightforward. Certainly the equation is of identical form to the associated Legendre equation, but clearly cannot have the same form of solution as we can see from the range of the variable: $+\infty \leqslant \xi \leqslant 1$. In fact a new function is introduced, the *associated Legendre function of the second kind*, $Q_l^m(\cosh \mu)$. No combination of choices of l and m will keep $Q_l^m(\xi)$ finite over the entire range of ξ however, and since stringent conditions are placed on l and m already in requiring finite, well-behaved solutions from equations (4.14.4), (4.14.5), all we can do is take whatever combination we can of the functions $P_l^m(\xi)$ and $Q_l^m(\xi)$ to keep $R(\xi)$ finite within the boundaries of the problem.

As a very simple example we may consider the potential distribution outside the prolate spheroidal surface ξ_0, given a surface distribution $\psi_0(\eta, \phi)$. The Laplace solution may be immediately written down as

$$\psi(\xi, \eta, \phi) = \sum_{l=0}^{\infty} \sum_{m=0}^{\infty} (A_{lm} \sin m\phi + B_{lm} \cos m\phi) P_l^m(\eta) Q_l^m(\xi)/Q_l^m(\xi_0)$$

(4.14.6)

and if we set $\psi_0(\xi_0, \eta, \phi) = $ constant, then we have $l = m = 0$, and

$$\psi(\xi) = \psi_0 \frac{Q_0^0(\xi)}{Q_0^0(\xi_0)}$$

(4.14.7)

that is, the equipotentials are confocal prolate spheroidal shells, and the flux lines are represented by the conjugate orthogonal confocal hyperboloids of revolution.

The functional form of the first few functions $Q_l^m(\xi)$ is given in Table 4.14.1 in both the ξ and hyperbolic notations. From it we are able to conclude our solution to the potential problem, and write equation (4.14.7) as

$$\psi(\xi) = \psi_0 \frac{\ln [(\xi + 1)/(\xi - 1)]}{\ln [(\xi_0 + 1)/(\xi_0 - 1)]}$$

(4.14.8)

The Helmholtz equation has its factored solutions, besides the usual trigonometric ϕ-solution, the solutions of the separated equations

$$\frac{d}{d\xi}\left[(\xi^2 - 1)\frac{dJ}{d\xi}\right] - \left[A - h^2\xi^2 + \frac{m^2}{\xi^2 - 1}\right]J = 0$$

(4.14.9)

$$\frac{d}{d\eta}\left[(1 - \eta^2)\frac{dS}{d\eta}\right] + \left[A - h^2\eta^2 - \frac{m^2}{1 - \eta^2}\right]S = 0$$

(4.14.10)

Table 4.14.1. Associated Legendre functions of the second kind $Q_l^m(\xi)$

	ξ-notation	Hyperbolic notation
Q_0^0	$\frac{1}{2}\ln\left(\dfrac{\xi+1}{\xi-1}\right)$	$\ln\left[\coth\left(\tfrac{1}{2}\mu\right)\right]$.
Q_1^0	$\frac{1}{2}\xi\ln\left(\dfrac{\xi+1}{\xi-1}\right)-1$	$\cosh\mu\ln\left[\coth\left(\tfrac{1}{2}\mu\right)\right]-1$
Q_1^1	$\sqrt{(\xi^2-1)}\left[\dfrac{\xi}{\xi^2-1}-\dfrac{1}{2}\ln\left(\dfrac{\xi+1}{\xi-1}\right)\right]$	$\coth\mu-\sinh\mu\ln\left[\coth\left(\tfrac{1}{2}\mu\right)\right]$
Q_2^0	$\frac{1}{4}(3\xi^2-1)\ln\left(\dfrac{\xi+1}{\xi-1}\right)-\dfrac{3}{2}\xi$	$\frac{1}{8}[3\cosh(2\mu)+1]\ln\left[\coth\left(\tfrac{1}{2}\mu\right)\right]-\dfrac{3}{2}\cosh\mu$
Q_2^1	$\sqrt{(\xi^2-1)}\left[\dfrac{3\xi^2-2}{\xi^2-1}-\dfrac{3}{2}\xi\ln\left(\dfrac{\xi+1}{\xi-1}\right)\right]$	$\operatorname{csch}\mu+3\sinh\mu-\dfrac{3}{2}\sinh(2\mu)\ln\left[\coth\left(\tfrac{1}{2}\mu\right)\right]$
Q_2^2	$\frac{3}{2}(\xi^2-1)\ln\left(\dfrac{\xi+1}{\xi-1}\right)-\dfrac{3\xi^3-5\xi}{\xi^2-1}$	$\frac{3}{2}[\cosh(2\mu)-1]\ln\left[\coth\left(\tfrac{1}{2}\mu\right)\right]$ $-\cosh\mu[3-2\operatorname{sech}^2\mu]$

$$\xi=\cosh\mu;\qquad \coth\left(\tfrac{1}{2}\mu\right)=\sqrt{\left(\dfrac{\xi+1}{\xi-1}\right)}$$

where, as in the elliptic case, $h=\frac{1}{2}ak=\pi a/\lambda$. Only for certain values of the separation constant A does equation (4.14.10) yield finite angular solutions, but for these discrete values it should not surprise us that the distribution may be conveniently expressed in terms of Legendre functions of the first kind, so that the angular spheroidal solution becomes:

$$S_{lm}(h,\eta)\Phi(m\phi)=\sum_{n=0}^{\infty}C(h,m,l)Y_{n+m}^m(\eta,\phi)\qquad(4.14.11)$$

Moreover, as the parameter $h\to 0$ (i.e. $a\to 0$) we regain the appropriate surface harmonic $S_{lm}(h,\eta)\to P_l^m(\eta)$ and $A\to l(l+1)$, as we would expect. There is a strict parallel here between the angular Mathieu functions in the elliptic case. Indeed, there is for the solution to the radial equation (4.14.10), where $J(\xi)$ is expressed as a series expansion of the spherical Bessel functions of the first and second kinds:

$$J_{lm}(h,\xi)=je_{lm}(h,\xi)+ye_{lm}(h,\xi)$$
$$=\left(\dfrac{\xi^2-1}{\xi^2}\right)^{m/2}\sum_{n}{}'D_n(h,l,m)\{j_{n+m}(h,\xi)+y_{n+m}(h,\xi)\}\qquad(4.14.12)$$

where the prime on the summation sign indicates inclusion of even ns when $(l-m)$ is even and odd ns when $(l-m)$ is odd. Again, as $h\to 0$ so we recover the limiting spherical solution $je_{l,m}\to j$.

Bates, Ledsham and Stewart (*Phil. Trans. Roy. Soc.* **A246**, 215, 1953) have applied this coordinate frame to the hydrogen molecular ion H_2^+. This is a stable system composed of two protons and a single electron, and the potential function of the system is

$$V = \frac{e^2}{R} - \frac{e^2}{r_1} - \frac{e^2}{r_2}$$

where R is the interproton (i.e. interfocal) separation and r_1, r_2 are the two proton–electron separations. If we assume that the two nuclei are stationary then Schrödinger's equation becomes

$$\nabla^2 \psi = \frac{2m}{\hbar^2} \left\{ E + \frac{e^2}{r_1} + \frac{e^2}{r_2} - \frac{e^2}{R} \right\} \psi = 0$$

where ∇^2 is the ellipsoidal operator (4.14.2) with η and ξ as defined in (4.14.1a). Separation of variables yields the usual three equations of the form (4.14.3), (4.14.4), and (4.14.5). Tables giving the solutions to these equations in terms of the functions discussed in this section are given by Bates *et al.* together with the eigenvalues E. Curves showing the energy as a function of internuclear separation are shown in Figure 4.14.2, and some of these evidently correspond to stable molecular systems. A much simpler approach to this system in terms of

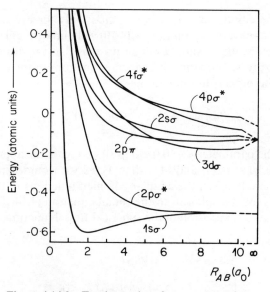

Figure 4.14.2 Total energies of some states of the He^+ molecule as a function of internuclear separation. Clearly some states represent bonding orbitals (cf. Figure 5.6.2). * denotes antibonding orbitals. (W. Kauzmann, *Quantum Chemistry*, 1957. Reproduced by permission of Academic Press, Inc.)

the variational approximation will be given in Section 5.6. It is instructive to compare the approximate curves of Figure 5.6.2 with the exact calculations shown in Figure 4.14.2.

If we rotate the confocal elliptic coordinates about the *minor* axis we obtain 'he oblate spheroidal coordinates. In these coordinates we have

$$
\nabla = \hat{\xi}\frac{1}{a}\sqrt{\frac{\xi^2 + 1}{\xi^2 + \eta^2}}\frac{\partial}{\partial \xi} + \hat{\eta}\frac{1}{a}\sqrt{\frac{1 - \eta}{\xi^2 - \eta^2}}\frac{\partial}{\partial \eta} + \hat{\phi}\frac{1}{a}\frac{1}{\sqrt{[(\xi^2 + 1)(1 - \eta^2)]}}\frac{\partial}{\partial \phi}
$$

(4.14.13)

and

$$
\nabla^2 = \frac{1}{a^2(\xi^2 + \eta^2)}\left\{\frac{\partial}{\partial \xi}(\xi^2 + 1)\frac{\partial}{\partial \xi} + \frac{\partial}{\partial \eta}(1 - \eta^2)\frac{\partial}{\partial \eta} + \frac{\xi^2 + 1 - 1 - \eta^2}{(\xi^2 + 1)(1 - \eta^2)}\frac{\partial^2}{\partial \phi^2}\right\}
$$

(4.14.14)

(cf. equations (4.14.1), (4.14.2)). As in the prolate spheroidal case, the 'radial' coordinate ξ goes from $0 \to \infty$ whilst η goes from -1 to $+1$.

The surfaces $\eta = $ constant are hyperboloids of one sheet asymptotic to the cone of semi-angle $\cos^{-1}\eta$, to the z-axis. The surfaces $\xi = $ constant (>0) are flattened spheroids. The separated Laplace equations are the same as for the prolate spheroids, equations (4.14.3), (4.14.4), (4.14.5), and the angular solutions are factored from

$$
\sin m\phi, \cos m\phi, P_l^m(\eta)
$$

whilst the radial solutions are the Legendre functions of the first and second kind, but of imaginary argument:

$$
P_l^m(i\xi), Q_l^m(i\xi)
$$

Problems

1 Determine the velocity potential at a point in an incompressible, irrotational fluid of density ρ due to a spherical bubble of radius a rising vertically with constant velocity v. Find the associated velocity distribution and calculate the kinetic energy of the system in the presence of the bubble. How is this accounted for physically?

2 A capacitor consists of concentric spheres of radii a and b. If the dielectric constant between the spheres varies as $1 + k/r$, where k is a constant, determine the capacitance of the system.

3 A uniform medium of dielectric constant ε has a small spherical cavity of radius a. An external field is applied which causes the medium at a large distance from the cavity to have a uniform field **E**. Determine the field in the cavity, and the charge distribution over its surface.

4 A superconductor may be regarded as a 'perfect diamagnetic' in that $B = 0$ inside it. Show that when a magnetized body is brought up to a plane superconductor the field is distorted as if there were a magnetic 'image' of the same polarity at the same distance behind the superconducting surface. Hence estimate the height at which a small rectangular bar magnet of mass m and magnetic moment m will float freely above the surface of a horizontal superconductor.

5 The study of nuclear energy levels may be approximated by a potential intermediate between the three-dimensional harmonic oscillator and the three-dimensional square-well potential. Correlate the energy levels in the two systems and discuss the degeneracy in the eigenvalues.

6 Calculate the first five absorption lines of the exciton in silicon. How many silicon atoms does the fifth orbit enclose?

7 Explain the significance of the wavefunction ψ which is used to describe an electron in quantum mechanics. What conditions regarding this quantity must be satisfied if the electron is in a stationary state?

An electron experiences a potential which is zero everywhere except for a spherical region of radius A, centred about a point O, within which it is uniformly attractive, say—V. Discuss the nature of ψ when the electron is in its state of lowest energy, for the limiting case when V is very large. What is the kinetic energy of this state, and what is the probability that an electron which occupies it will be found near O, within a spherical region of radius a $(\ll A)$?

Explain why if $V < \hbar^2/mA^2$ it is impossible for an electron of mass m to be bound in the neighbourhood of O, unless its angular momentum about O is zero.

(Detailed solution of Schrödinger's equation is not required.)

8 What do you understand by the Heisenberg uncertainty principle? Illustrate your answer as quantitatively as you can by reference to the following phenomena: (a) the diffraction of waves, (b) the finite width of spectral lines, (c) the zero point energy of harmonic oscillators.

9 Prove Gauss' theorem and discuss its application to electrostatics, magnetism and gravitation.

Use it to prove that the electrostatic field inside a hollow conductor that contains no charge is zero, whatever the shape of the conductor and whatever charges are present in the neighbourhood of (but external to) the conductor.

Why is it not usually true that the gravitational field inside a thin shell of matter is zero? What condition would have to be satisfied for it to be true for an ellipsoidal shell of matter isolated in space? (Only a qualitative answer is required.)

10 Show that the field inside a uniform dielectric sphere placed in a uniform field is itself uniform.

It is desired to create an electric field of 10 volt cm^{-1} inside a dielectric sphere of dielectric constant 2·5. What externally applied field is necessary?

11 The Schrödinger equation for a hydrogen atom is

$$\nabla^2\psi + \frac{2m}{\hbar^2}\left(E + \frac{e^2}{r}\right)\psi = 0$$

Verify that the normalized wavefunction for the ground state can be written as

$$\psi_0 = K \exp(-r/a)$$

and, hence, find expressions for the energy of the ground state and the constants a and K.

Show that the probability of finding the electron further from the proton than a distance d is approximately given by the expression

$$2(d/a)^2 \exp(-2d/a) \quad \text{for } d \gg a$$

(For spherical symmetry,

$$\nabla^2 \psi = \frac{1}{r^2} \frac{\partial}{\partial r}\left(r^2 \frac{\partial \psi}{\partial r}\right)\bigg)$$

12 When slow electrons are passed through a monatomic gas, it is found that for certain electron energies the gas becomes abnormally 'transparent' and scatters the electrons only slightly (Ramsauer effect). This effect can be qualitatively understood in terms of a one-dimensional model, in which the electron moves through a potential $V = 0$ for $x < 0$ and $x > a$, and a potential $V = V_0$ for $0 \leqslant x \leqslant a$ (with V_0 negative). Show that this potential well becomes transparent if the wavelength of the electron in the well is $2a/n$, where n is an integer. If $a = 3 \times 10^{-8}$ cm and $V_0 = -4$ eV, calculate the lowest electron energy E for which the well becomes transparent.

13 The wave equation for a particle placed in a weak magnetic field H in the z-direction is

$$\nabla^2 \psi + \frac{2m}{\hbar^2}\left\{E - V(r) + \mu H \frac{1}{i}\frac{\partial}{\partial \phi}\right\}\psi = 0$$

where μ is the Bohr magneton.

If $V(r)$ is a spherical potential well with infinite sides, i.e.

$$V(r) = 0 \quad (r < a)$$
$$V(r) = \infty \quad (r > a)$$

show that a possible wavefunction for the particle is

$$\psi = \text{const}\left(\frac{\sin \alpha r}{(\alpha r)^2} - \frac{\cos \alpha r}{\alpha r}\right)\sin \theta \, e^{i\phi}$$

and determine the value of α and the eigenenergy. Identify the state of motion of the particle.

$$\nabla^2 = \left\{\frac{1}{r^2}\frac{\partial}{\partial r}\left(r^2 \frac{\partial}{\partial r}\right) + \frac{1}{r^2 \sin \theta}\frac{\partial}{\partial \theta}\left(\sin \theta \frac{\partial}{\partial \theta}\right) + \frac{1}{r^2 \sin^2 \theta}\frac{\partial^2}{\partial \phi^2}\right\}$$

14 A function of position is given in terms of spherical polar coordinates by $\psi = r^n \cos \theta$. Evaluate $\nabla^2 \psi$ and find the values of n for which $\nabla^2 \psi = 0$, $\propto r \cos \theta$.

A spherical body of radius a has density $\lambda \zeta$ where λ is a constant and ζ is the distance from a tangent plane. Find the associated gravitational potential everywhere.

CHAPTER 5

Approximate Methods and Applications

5.1 Introduction

As we have observed at various stages throughout this book, there is no guarantee that the differential field equation will separate nor, should it separate, is it by any means certain that the resulting equations will be solvable. The class of Helmholtz equations will generally separate in the eleven coordinate frames obtainable from the generalized ellipsoidal coordinates, but suffers from rapid mathematical intractability as soon as the slightest modification in the form of the boundary conditions or the coordinate frame itself is introduced. The Schrödinger equation is even more susceptible in this respect since only for certain functional forms of the Hamiltonian is separation possible, and the number of such cases is strictly limited.

Quite clearly we require some systematic means of obtaining approximate solutions in these cases when no exact solution is possible, preferably with some indication of the associated error incurred in making the approximation. It is for this reason that we come now to consider three of the principal approximate methods of solution of the differential field equations outlined in this book. The modification of the boundary conditions or distortion of the coordinate frame may be small with respect to an exactly solvable problem in which case a *perturbation* approach is implied, taking the unperturbed system as the zeroth-order approximation to the perturbed system. Thus the perturbed Hamiltonian $\mathscr{H}$ may be written as

$$\mathscr{H} = \mathscr{H}^{(0)} + \lambda \mathscr{H}^{(1)} \tag{5.1.1}$$

where $\lambda \ll 1$ is a measure of the strength of the perturbation, and is separated from the zeroth order part $\mathscr{H}^{(1)}$ for the purpose of determining infinitesimals in power series expansions. The perturbed solutions ψ_n may, for small perturbations, be Fourier expanded in terms of a complete orthonormal set of unperturbed functions in the usual way

$$\psi_n = \psi_n^{(0)} + \sum_{n \neq i} a_{ni} \psi_i^{(0)} \tag{5.1.2}$$

where the coefficients $a_{ni} \ll 1$, and are functions of λ. The zero-order function $\psi_n^{(0)}$ dominates, and as $\lambda \to 0$ so the unperturbed solution is regained.

In the solution of the Schrödinger equation in particular, although the method applies equally to the Helmholtz equation, we may well encounter potential functions $V(r)$ for which there is no exact solution, nor may $V(r)$ be considered as a perturbation of any exactly solvable potential. Under these circumstances, provided $V(r)$ is slowly varying spatially the Wentzel–Kramers–Brillouin (WKB) method may apply. In this the *local* solution is established on the assumption that the functional form of the potential has a *constant* average value over that region and does not have a spatial coordinate dependence. The solutions are subsequently matched at the boundaries of the local regions. Clearly this method cannot apply in regions where $V(r)$ is a rapidly varying function of position.

The third approximate method we shall consider is the *variational approximation* which applies when the perturbation becomes large. In this approach a parametrized trial function is inserted for the unknown distribution $\psi(\mathbf{r})$, and then some physical functional $Q(\psi_{\text{trial}})$, frequently the free energy, is calculated, Q being stationary upon variation of the function, i.e. $(\partial Q/\partial\psi_{\text{trial}}) = 0$. The trial function may be varied by changing the values of the parameters. The crux of this method lies in the initial choice of trial function which should satisfy three conditions: (i) it must satisfy the boundary conditions, (ii) it should have the correct qualitative form of the expected distribution, and (iii) it should lead to relatively simple functionals $Q(\psi_{\text{trial}})$ so that the functional differential $(\partial\phi/\partial\psi_{\text{trial}})$ may be determined. The initial trial function is therefore chosen as a compromise between qualitative physical correctness and mathematical expediency, and is often based purely upon physical intuition. Guidance regarding the initial choice may be given by the Fourier representation of the distribution in terms of an appropriate complete orthonormal set. Unlike the perturbation treatment, however, no zeroth-order eigenfunction will dominate (as in equation (5.1.2)): the Fourier coefficients a_i will all be either of the same order of magnitude, or zero.

5.2 Perturbation Theory for Non-Degenerate States

As we observed in the previous section and in Sections 2.3 and 4.12, the perturbation approach is particularly appropriate in situations which closely resemble the solvable one, and the implicit assumption is made that the perturbed situation is attainable from the unperturbed condition by means of a continuous variation of a parameter λ. λ serves to 'switch on' the perturbation and as such is a measure of the 'strength' of the perturbation.

We generally attempt to expand perturbed functions as a series in λ, and as $\lambda \to 0$ so we regain the unperturbed or zero-order function. It may be necessary to consider second-order or third-order perturbations, that is, terms of order λ^2, λ^3—but it is nevertheless assumed that there is a radius of convergence of this series in λ within which we may obtain an accurate representation of the perturbed function. It is possible, incidentally, to make an analytic continuation of the series beyond the radius of convergence if necessary, but that will not concern us here.

We have mentioned (Section 1.7) that it is possible to reduce all the field equations encountered in this book to Sturm–Liouville form. That is, we may generally write (equation (1.7.2))

$$\frac{d}{dx}\left\{u(x)\frac{d\psi_n(x)}{dx}\right\} - h_1(x)\psi_n(x) = k_n^2 h_2(x)\psi_n(x) \tag{5.2.1}$$

where $h_2(x)$ is termed the weighting function, and generally depends upon the system of coordinates, and k_n^2 is the appropriate eigenvalue or separation constant. Equation (5.2.1) may be particularly simply expressed as follows, where the superscript $^{(0)}$ represents an explicitly unperturbed system:

$$\mathscr{L}^{(0)}\psi_n^{(0)}(x) = k_n^{2(0)}h^{(0)}(x)\psi_n^{(0)}(x) \tag{5.2.2}$$

$\mathscr{L}^{(0)}$ is known as the unperturbed Sturm–Liouville operator

$$\mathscr{L}^{(0)} = \frac{d}{dx}\left\{u(x)\frac{d}{dx}\right\} - h_1(x) \tag{5.2.3}$$

and the weighting function has been written $h^{(0)}(x)$. Now, under the effect of a perturbation the Sturm–Liouville operator will modify thus $\mathscr{L} \rightarrow \mathscr{L}^{(0)} + \lambda\mathscr{L}^{(1)}$, and similarly the weighting function $h \rightarrow h^{(0)} + \lambda h^{(1)}$. Equation (5.2.2) then reads

$$(\mathscr{L}^{(0)} + \lambda\mathscr{L}^{(1)})\psi_n = k_n^2(h^{(0)} + \lambda h^{(1)})\psi_n \tag{5.2.4}$$

If we now write the *perturbed* functions ψ_n, k_n^2 as a power series in λ, thus

$$\psi_n = \psi_n^{(0)} + \lambda\psi_n^{(1)} + \lambda^2\psi_n^{(2)} + \cdots \tag{5.2.5}$$

$$k_n^2 = k_n^{2(0)} + \lambda k_n^{2(1)} + \lambda^2 k_n^{2(2)} + \cdots \tag{5.2.6}$$

where the superscripts $^{(0), (1), (2)},\ldots$, represent the zeroth-, first-, second-,$\ldots$, order perturbations, then for small values of λ, i.e. small perturbations, the series rapidly converges. The problem then becomes one of finding the functional form of $\psi_n^{(1)}, \psi_n^{(2)}, \ldots$ and $k_n^{2(1)}, k_n^{2(2)}, \ldots$, it being understood that the zero-order terms $^{(0)}$ represent the unperturbed functions throughout.

We are by now familiar with the approach to be made: the functions $\psi_n^{(1)}, \psi_n^{(2)}, \ldots$, are expressed in terms of a complete set of orthonormal unperturbed functions thus:

$$\psi_n^{(1)} = \sum_i A_{ni}\psi_i^{(0)}, \tag{5.2.7}$$

$$\psi_n^{(2)} = \sum_i B_{ni}\psi_i^{(0)}, \quad \text{etc.} \tag{5.2.8}$$

so now the problem is reduced to determining the coefficients $A_{ni}, B_{ni}, \ldots$, together with the functions $k_n^{2(1)}, k_n^{2(2)}, \ldots$. Substituting equations (5.2.5)–(5.2.8) in equation (5.2.4) we obtain

$$(\mathscr{L}^{(0)}\psi_n^{(0)} - k_n^{2(0)}h^{(0)}\psi_n^{(0)})$$

$$+ \left\{\sum_i A_{ni}(k_i^{2(0)} - k_n^{2(0)})h^{(0)}\psi_i^{(0)} + (\mathscr{L} - k_n^{2(0)}h)\psi_n^{(0)} - k_n^{2(1)}h^{(0)}\psi_n^{(0)}\right\}\lambda$$

$$+ \left\{\sum_i B_{ni}(k_i^{2(0)} - k_n^{2(0)})h^{(0)}\psi_i^{(0)} + \sum_i A_{ni}(\mathscr{L} - k_n^{2(0)}h - k_n^{2(1)}h^{(0)})\psi_{ni}^{(0)}\right.$$

$$\left. - (k_n^{2(2)}h^0 + k_n^{2(1)}h)\psi_n^{(0)}\right\}\lambda^2 \tag{5.2.9}$$

$$+ \cdots = 0$$

The first term in the series is zero, from equation (5.2.2). Comparing coefficients in $\lambda, \lambda^2, \ldots$, we conclude that the expressions in the curly brackets must vanish. Thus, from the coefficient of λ

$$\left\{\sum_i A_{ni}(k_i^{2(0)} - k_n^{2(0)})h^{(0)}\psi_i^{(0)} + (\mathscr{L} - k_n^{2(0)}h)\psi_n^{(0)} - k_n^{2(1)}h^{(0)}\psi_n^{(0)}\right\} = 0 \tag{5.2.10}$$

Multiplying throughout by $\psi_n^{(0)}$ and integrating over a suitable range, the orthogonality of the unperturbed eigenfunctions yields

$$k_n^{2(1)} = \int_a^b \psi_n^{(0)}(\mathscr{L} - k_n^{2(0)}h)\psi_n^{(0)}\,\mathrm{d}x \tag{5.2.11}$$

which represents the *first-order perturbation of the nth eigenvalue*.

If we now multiply throughout by $\psi_i^{(0)}$ and integrate we obtain the *first-order perturbation of the nth eigenfunction*

$$\left. \begin{aligned} A_{ni} &= \frac{1}{k_n^{2(0)} - k_i^{2(0)}} \int_a^b \psi_i^{(0)}(\mathscr{L} - k_n^{2(0)}h)\psi_n^{(0)}\,\mathrm{d}x \\[2mm] &\equiv \frac{\mathscr{L}_{in}}{k_n^{2(0)} - k_i^{2(0)}} \quad \text{where } i \neq n \end{aligned} \right\} \tag{5.2.12}$$

Similarly, from the coefficient of λ^2 we determine the *second-order perturbation of the nth eigenvalue* as

$$k_n^{2(2)} = \sum_{i \neq n} \frac{\mathscr{L}_{ni}(\mathscr{L}_{ni} - k_n^{2(0)}h_{ni})}{k_n^{2(0)} - k_i^{2(0)}} - \mathscr{L}_{nn}h_{nn} + \tfrac{1}{2}k_n^{2(0)}h_{nn}^2 \tag{5.2.13}$$

where

$$h_{ni} = \int_a^b \psi_i^{(0)}h\psi_n^{(0)}\,\mathrm{d}x,$$

and for the *second-order perturbation of the nth eigenfunction*

$$\begin{aligned} B_i &= \sum_{j \neq n} \frac{\mathscr{L}_{nj}(\mathscr{L}_{nj} - k_n^{2(0)}h_{ij})}{(k_n^{2(0)} - k_j^{2(0)})(k_n^{2(0)} - k_i^{2(0)})} - \frac{\mathscr{L}_{in}\mathscr{L}_{nn}}{(k_n^{2(0)} - k_i^{2(0)})^2} \\[2mm] &\quad - \frac{\mathscr{L}_{nn}h_{in} + \tfrac{1}{2}\mathscr{L}_{in}h_{nn} - \tfrac{1}{2}k_n^{2(0)}h_{nn}h_{ni}}{k_n^{2(0)} - k_i^{2(0)}} \quad \text{where } i \neq n \end{aligned} \tag{5.2.14}$$

and so on.

This treatment is seen to be a generalization of the hybridization concept dealt with at some length in Section 2.3. As we observed at the time, the coefficients $A_{ni}, B_{ni}, \ldots$, represent the 'contamination' of the nth unperturbed eigenfunction by the ith: that is, to first order, A_{ni} is the coefficient of the $\psi_n - \psi_i$ hybrid. Again, we mentioned that *adjacent* eigenfunctions seem to hybridize most strongly, that is, those for which $k_n^{2(0)} \sim k_i^{2(0)}$. Alternatively we may say those distributions having similar natural 'frequencies' hybridize most extensively. In the degenerate case, when we have several eigenfunctions having the same eigenvalue, hybridization occurs even in the absence of a perturbation. Clearly, the perturbation theory as developed here cannot apply to degenerate eigenfunctions since we would in this case experience difficulty with vanishing denominators in expressions such as (5.2.12) and (5.2.13). As examples of this method we direct the reader to Sections 2.3 and 4.12.

5.3 Perturbation Theory for Degenerate States

The perturbation theory outlined in the previous section requires some modification before it may be applied to degenerate states. In the non-degenerate case the zeroth-order eigenfunction may be unequivocally specified. In the degenerate case, however, the zeroth-order eigenfunction will in general be a hybridized linear combination of the degenerate eigenfunctions. Thus, if the eigenfunction is l-fold degenerate the zeroth-order (unperturbed) degenerate wavefunction will be written as the hybrid

$$\hat{\psi}_n^{(0)} = \sum_{j=1}^{l} c_{jj}\psi_n^{(0)} \tag{5.3.1}$$

where $\hat{}$ signifies a degenerate state and $_j\psi_n^{(0)}$ represents the jth of the l unperturbed eigenfunctions in the nth energy state. We now have to determine the coefficients c_j in addition to the $A_i, B_i, \ldots, k_n^{2(1)}, k_n^{2(2)}, \ldots$, discussed in the previous section.

The nth *perturbed* degenerate eigenfunction may now be written

$$\hat{\psi}_n = \sum_{j=1}^{l} c_{j}\,_j\psi_n^{(0)} + \lambda \sum_{i} A_{ni}\psi_i^{(0)} + \cdots \tag{5.3.2}$$

$$k_n^2 = k_n^{2(0)} + \lambda k_n^{2(1)} + \cdots \tag{5.3.3}$$

These expressions should be compared with the corresponding non-degenerate expansions (5.2.6), (5.2.7). We see that they differ only in the hybrid expression of the zeroth-order eigenfunction.

The perturbed Sturm–Liouville equation now becomes (cf. equation (5.2.4))

$$(\mathcal{L}^{(0)} + \lambda\mathcal{L}^{(1)})\hat{\psi}_n = k_n^2(h^{(0)} + \lambda h^{(1)})\hat{\psi}_n \tag{5.3.4}$$

Following the same procedure as in the non-degenerate case, equating powers of λ, and making use of the unperturbed equation, we have

$$(k_i^{2(0)} - k_n^{2(0)})h^{(0)}A_{ni}\,_j\psi_i^{(0)} + \mathcal{L}\sum_{j}^{l} c_{j}\,_j\psi_n^{(0)} - (k_n^{2(0)}h - k_n^{2(1)}h^{(0)})\sum_{j}^{l} c_{j}\,_j\psi_n^{(0)} = 0 \tag{5.3.5}$$

Multiplying equation (5.3.5) throughout by one of the l degenerate eigenfunctions, say $_l\psi_n^{(0)}$, and integrating over the space variables we obtain

$$\sum_j^l c_j(\mathscr{L}_{lj} - \delta_{lj}k_n^{2(1)}) = 0; \quad \delta_{lj} \begin{cases} =0 & l \neq j \\ =1 & l = j \end{cases}$$

(5.3.6)

where the simplification has arisen on account of the orthogonality of the eigenfunctions. If we repeat this for each of the degenerate eigenfunctions $_1\psi_n^{(0)}, _2\psi_n^{(0)}, \ldots, _l\psi_n^{(0)}$, we shall obtain l linear equations of the same form as equation (5.3.6). This set of homogeneous linear equations may have a set of non-vanishing solutions $c_1, \ldots, c_j$ provided the determinant of coefficients vanishes. That is, from equation (5.3.6),

$$\begin{vmatrix} (\mathscr{L}_{11} - k_n^{2(1)}) & \mathscr{L}_{12} & \cdots & \mathscr{L}_{1l} \\ \mathscr{L}_{21} & (\mathscr{L}_{22} - k_n^{2(1)}) & \cdots & \mathscr{L}_{2l} \\ \vdots & & & \vdots \\ \mathscr{L}_{l1} & \mathscr{L}_{l2} & \cdots & (\mathscr{L}_{ll} - k_n^{2(1)}) \end{vmatrix}$$

(5.3.7)

This equation gives the condition which the first-order perturbation $k_n^{2(1)}$ has to satisfy. The only unknown in this determinant is $k_n^{2(1)}$, and the lth-order algebraic equation will have l roots, $k_{n1}^{2(1)}, \ldots, k_{nl}^{2(1)}$. Each root of the determinant (5.3.7) gives rise to a set of solutions to equation (5.3.6) for c_j. This determines the zero-order degenerate eigenfunction $\hat{\psi}_n^{(0)}$. That the initially degenerate wavefunction now has l eigenfunctions and eigenvalues shows that the degeneracy has been eliminated by the application of a perturbation. If some of the roots remain coincident, then the degeneracy has not been entirely removed, although it may be eliminated at second order. It can be shown that the l roots of the determinant (5.3.7) may be expressed in terms of the *set* of l unperturbed eigenfunctions, $\hat{\psi}_n^{(0)}$,

$$k_n^{2(1)} = \int_a^b \hat{\psi}_n^{(0)}(\mathscr{L} - k_n^{2(0)}h)\hat{\psi}_n^{(0)} \, dx$$

(5.3.8)

which should be compared to equation (5.2.12) in the non-degenerate case.

Having determined the values of $k_n^{2(1)}$ and c_j, we must now determine the A_{ni} in equation (5.3.2). These are determined as in the non-degenerate case by multiplying equation (5.3.5) throughout by $\psi_i(0)$ and integrating over the space variables:

$$A_{ni} = \frac{1}{k_n^{2(0)} - k_i^{2(0)}} \int_a^b \psi_i^{(0)}(\mathscr{L} - k_n^{2(0)}h)\left(\sum_j^l c_j \, _j\psi_n^{(0)}\right) dx \quad \text{where } i \neq n$$

(5.3.9)

which is of the same form as equation (5.2.12). This is, of course, just what we should expect since once the degeneracy has been removed by the perturbation each zeroth-order eigenfunction acts like a non-degenerate eigenfunction.

5.4 The WBKJ Approximation

Any linear homogeneous second-order differential equation may be written in the form

$$\frac{d^2\psi(x)}{dx^2} + f(x)\psi(x) = 0 \qquad (5.4.1)$$

and Wentzel, Brillouin and Kramers independently established a general approximate form of solution subject to $f(x)$ being a slowly varying function: we shall be more precise about this point shortly. The method had also been given earlier by Jefferys. The method is important in that it enables us to obtain approximate solutions to the Schrödinger equation for arbitrary, but slowly varying, forms of potential $V(x)$. It is also of more general applicability.

The essence of the method is to observe that if $f(x)$ were a constant, f, then we should immediately have solutions $\psi(x)$ of the form $\sin\sqrt{f}x$ or $\cos\sqrt{f}x$ if $f < 0$, and $\exp \pm\sqrt{f}x$ if $f > 0$. Now if $f(x)$ is no longer constant, but instead a slowly varying function, it might be reasonable to assume that the solution would not be markedly different and would be associated somehow with a local *average* value of $f(x)$.

If we assume a form of solution

$$\psi = e^{i\phi(x)} \qquad (5.4.2)$$

equation (5.4.1) becomes

$$-(\phi'(x))^2 + i\phi''(x) + f(x) = 0 \qquad (5.4.3)$$

If we now assume $\phi''(x)$ to be small in comparison to the other quantities, which it will be for a slowly varying function, we have

$$\phi'(x) = \pm\sqrt{f(x)} \qquad (5.4.4)$$

i.e.

$$\phi(x) = \pm \int^x \sqrt{f(x)}\,dx$$

which represents the first order of an iterated solution. If we now write, from equation (5.4.4),

$$\phi''(x) = \pm\tfrac{1}{2}f'(x)(f(x))^{-\frac{1}{2}} \qquad (5.4.5)$$

and substitute in equation (5.4.3) we obtain

$$(\phi'(x))^2 = f(x) \pm \frac{i}{2}\frac{f'(x)}{\sqrt{f(x)}}$$

$$\phi'(x) = \pm \sqrt{f(x)} + \frac{i}{4}\frac{f'(x)}{f(x)}$$

$$\phi(x) = \pm \int^x \sqrt{f(x)}\,dx + \frac{i}{4}\log_e f(x) \qquad (5.4.6)$$

When $f(x) > 0$ the general solution of equation (5.4.1) takes the form

$$\psi_{\mathrm{I}} = \frac{A}{\sqrt[4]{f(x)}} \sin\left(\int^x \sqrt[4]{f(x)}\,dx\right) + \frac{B}{\sqrt[4]{f(x)}} \cos\left(\int^x \sqrt[4]{f(x)}\,dx\right) \quad (5.4.7)$$

where the positive root is taken throughout and where A and B are arbitrary constants. When $f(x) < 0$ the general solution may be written as

$$\psi_{\mathrm{II}} = \frac{C\,e^{\int^x \sqrt{-f(x)}\,dx}}{\sqrt[4]{-f(x)}} + \frac{D\,e^{-\int^x \sqrt{-f(x)}\,dx}}{\sqrt[4]{-f(x)}} \quad (5.4.8)$$

where C and D are arbitrary constants. Again, the positive roots are taken.

In Figure 5.4.1 we indicate the general forms of solution: ψ_{I} in the interval $a < x < b$ (equation (5.4.7)) and ψ_{II} in the intervals $a > x > b$ (equation (5.4.8)). Two points in particular arise in the vicinity of the singular points $x = a, b$

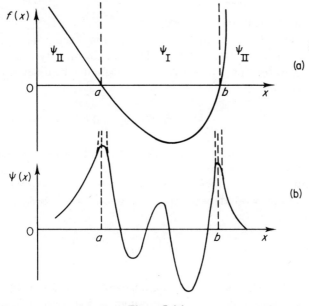

Figure 5.4.1

when $f(x) = 0$ and both ψ_{I} and ψ_{II} become infinite. We have first of all to establish the form of solution at these points, and secondly to ensure that they match to form a continuous solution. The problem of singularities at $x = a, b$ is overcome by approximating $f(x)$ to a straight line in these regions, and then to shift the origin to a or b. $f(x)$ then takes the form $f(x) = c^2(x - a, b)$ where c^2 is a constant, and solutions to equation (5.4.1) are subsequently given in terms of Bessel functions: this point will be discussed further in a moment.

The second problem of connecting the exponential to the oscillatory type of solution is difficult, and so-called *connection formulae* have been established relating A, B and C, D on either side of the points $f(x) = 0$. We shall not establish the connection here, but it involves avoiding the points a, b on the real axis and encircling them in the complex plane along a path for which the WKB approximation remains valid.

If the function $f(x)$ passes through zero at the two points a, b only, and since we are interested in solutions which are well-behaved at infinity, only the decaying exponential functions in the regions $a > x > b$ are acceptable. Under these circumstances the connection formulae become

$$
\underset{\psi_{\text{II}}}{\underbrace{\frac{\alpha\, e^{-\int_{x}^{a}\sqrt{-f(x)}\,dx}}{\sqrt[4]{-f(x)}}}} \leftrightarrow \underset{\psi_{\text{I}}}{\underbrace{\frac{2\alpha}{\sqrt[4]{f(x)}}\cos\left(\int_{a}^{x}\sqrt{f(x)}\,dx - \frac{\pi}{4}\right)}}
$$

$$
\frac{2\beta}{\sqrt[4]{f(x)}}\cos\left(\int_{x}^{b}\sqrt{f(x)}\,dx - \frac{\pi}{4}\right) \leftrightarrow \underset{\psi_{\text{II}}}{\underbrace{\frac{\beta\, e^{-\int_{b}^{x}\sqrt{-f(x)}\,dx}}{\sqrt[4]{-f(x)}}}} \tag{5.4.9}
$$

where α and β are arbitrary constants.

As we mentioned earlier, in the vicinity of the singularity at $x = a, b$ we replace $f(x)$ by a linear extrapolation. Thus, in the vicinity of $x = a$ we may approximate $f(x)$ by $c^2(x - a)$ where c is a constant. The differential equation (5.4.1) now becomes

$$
\frac{d^2\psi}{dx^2} + c^2(x - a)\psi = 0
$$

in the vicinity of the singularity, (making the substitution $z = -c^{\frac{2}{3}}(x - a)$ transforms the above equation into *Airy's equation*; $\psi'' - z\psi = 0$). An exact solution of equation (5.4.1) may then be found which takes the form of a linear combination of the two independent Bessel functions of $\frac{1}{3}$-order: $f^{-\frac{1}{2}}\xi^{\frac{1}{2}}J_{\pm\frac{1}{3}}(\xi)$ where

$$
\xi = \int_{a}^{x} f\, dx
$$

in the vicinity of a. This represents an approximate solution to the differential equation at $x = a$ and at $x = \pm\infty$, although it does depart from the exact solution elsewhere. The solution is, in fact, very closely related to the Airy function. In the region for which $f^2 > 0$ the Bessel functions tend asymptotically to trigonometric functions, and the solutions can therefore be matched with those in the interval $a < x < b$ shown in equation (5.4.9). On the other hand, in the regions for which $f^2 < 0$ we may formally replace the argument f by $i|f|$ whereupon we obtain the Bessel functions of imaginary argument, of which we require the real component—a non-oscillatory spatially decaying function (See Sections 3.5 and 3.6). We may therefore asymptotically match the solutions in the two regions around the singularity.

By way of example we may consider the Schrödinger equation for which $f(x) = 2m\hbar[E - V(x)]$ (Figure 2.4.1a). It is quite clear that the points $x = a, b$

represent those points at which $E = V(x)$, and for this reason are known as the *classical turning points*.

Reference to equation (5.4.9) shows that the two connection formulae in the region $x < b, x > a$ must be identical:

$$\frac{2\alpha}{\sqrt[4]{(E - V)}} \cos \left\{ \int_a^x \sqrt{\left[\frac{2m}{\hbar^2}(E - V)\right]} \, dx - \frac{\pi}{4} \right\}$$

$$= \frac{2\beta}{\sqrt[4]{(E - V)}} \cos \left\{ \int_x^b \sqrt{\left[\frac{2m}{\hbar^2}(E - V)\right]} \, dx - \frac{\pi}{4} \right\} \qquad (5.4.10)$$

from which we conclude

$$\int_a^b \sqrt{[2m(E - V)]} \, dx = (n + \tfrac{1}{2})n\hbar \qquad (5.4.11)$$

This expression, we notice, is closely related to the Wilson–Sommerfeld quantum condition $\oint p \, dq = nh$ where p and q are the generalized momentum and position coordinates respectively. Equation (5.4.11) is, therefore, the condition for obtaining the energy eigenvalues in the general one-dimensional case.

Returning to equation (5.4.3) we see that for the validity of the WKB approximation the 'curvature' ϕ'' is to be negligible in comparison to f. That is, from equation (5.4.5)

$$\phi'' = \pm \frac{1}{2} \left| \frac{f'}{\sqrt{f}} \right| \ll f \qquad (5.4.12)$$

Thus from equations (5.4.2) and (5.4.4) we see that $f^{-\frac{1}{2}}$ is of the order of 2π times one wavelength or characteristic exponential length of the solution ψ. The condition evidently amounts to the fact that $f(x)$ should not change significantly within one wavelength; this condition is satisfied in particular in the high quantum number region where the wavelength is small.

Inspection of the expressions for the eigenfunctions in the region $a < x < b$ (equation (5.4.9)) shows that the amplitude and wavelength in the high quantum number region is governed by the 'potential function' $f(x)$. Thus the amplitude of the cosine function is modulated as $1/\sqrt[4]{f(x)}$, whilst the wavelength varies as $2\pi/\sqrt{f(x)}$. This is particularly well shown in the case of the harmonic oscillator: that this is the case at high quantum numbers (and incidentally is *not* the case at low quantum numbers) is further evidence of Bohr's correspondence principle.

5.5 The Variational Approximation

The number of differential field equations which will actually succumb to a straightforward separation of variables and subsequent exact solution in terms of the well known functions is very small indeed. As we have observed, the field equations have an unfortunate tendency of becoming distinctly intractable at the slightest opportunity, although the general perturbation theory outlined in Sections 5.2 and 5.3 goes some of the way towards an estimate of the first- and second-order corrections to be applied in cases not too far removed from the solvable situation. However, situations arise in which neither the perturbation theory nor the WKB approach will enable us to obtain physically satisfactory approximate solutions to the field equations.

The variational method to be discussed here is based on a simple theorem according to which the eigenvalue of the approximate solution will always be greater than, or at best equal to, the *actual* eigenvalue. Choosing a parametrized trial solution ψ(trial), the optimum parametric form of this trial solution may be determined by finding which values of the coefficients yield the lowest eigenvalue. This, according to the variational theorem, must be the optimum form *for that choice of trial function*. If a completely different form of trial function were chosen, and it was parametrically optimized to yield its lowest eigenvalue, this being lower than the previous estimate, then clearly the latter trial function is better.

First, however, we shall consider the variational theorem applied to the general Sturm–Liouville equation. The Sturm–Liouville differential field equation (5.2.3)

$$\mathscr{L}\psi(x) = k^2 h(x)\psi(x) \tag{5.5.1}$$

will have as solutions satisfying the boundary conditions at $x = a, b$ the set of eigenfunctions $\psi_0(x), \psi_1(x), \psi_2(x), \ldots, \psi_n(x)$, with associated eigenvalues k_0^2, $k_1^2, \ldots, k_n^2$. Suppose we now propose the *trial solution* $\tilde{\psi}(x)$ as the solution to equation (5.5.1). $\tilde{\psi}(x)$ must, necessarily, satisfy the boundary conditions at $x = a, b$. We may, of course, express the trial function in terms of the complete set as follows:

$$\tilde{\psi}(x) = \sum_{n=0}^{\infty} A_n \psi_n(x) \tag{5.5.2}$$

whereupon

$$\mathscr{L}\tilde{\psi}(x) = \sum_{n=0}^{\infty} A_n \mathscr{L}\psi_n(x) = \sum_{n=0}^{\infty} A_n k_n^2 h(x)\psi(x) \tag{5.5.3}$$

If we now define a quantity

$$Q = \frac{\int_a^b \tilde{\psi}\mathscr{L}\tilde{\psi} \, dx}{\int_a^b \tilde{\psi}\tilde{\psi} h \, dx} \tag{5.5.4}$$

then from equations (5.5.2) and (5.5.3) we have immediately

$$Q = \frac{\sum\sum A_n A_m k_n^2 \int_a^b \psi_n \psi_m h \, dx}{\sum\sum A_n A_m \int_a^b \psi_n \psi_m h \, dx} = \frac{\sum A_n^2 k_n^2}{\sum A_n^2} \tag{5.5.5}$$

So that we may further write

$$Q - k_0^2 = \frac{\sum A_n^2 (k_n^2 - k_0^2)}{\sum A_n^2} \tag{5.5.6}$$

where k_0^2 is numerically the lowest eigenvalue. Since the A_n are all positive it follows immediately that

$$Q \geqslant k_0^2$$

Moreover, if $Q = k_0^2$, the only way the right-hand side of equation (5.5.6) can vanish is for all $A_n = 0$, except for A_0. This being the case, it follows from equation (5.5.2) that the optimum form of the trial function is

$$\tilde{\psi}(x) = A_0 \psi(x) \qquad (5.5.7)$$

that is, the *exact* form of eigenfunction has the lowest Q-value. This, then, provides us with a means for converging on the optimum form of trial function, although we can never know whether it is the exact function. Q, in fact, may be identified as the free energy of the system, and we are effectively finding that distribution $\tilde{\psi}$ which minimizes the free energy.

It is rather ineffective to guess at isolated $\tilde{\psi}$s, and is unnecessarily tedious. We can deal at once with a whole family of $\tilde{\psi}$ by choosing a trial function with a number of variable parameters $a_1, a_2, \ldots, a_n$. The values of these parameters finally chosen are those which minimize Q. Obviously the more parameters there are in the trial function the more flexible it is, and generally we should expect to achieve lower and lower eigenvalues. The particular parametric form of $\tilde{\psi}$ which achieves the lowest Q-value represents the optimum form *of that particular trial function*. In Figure 5.5.1 we show how the Q-value will vary

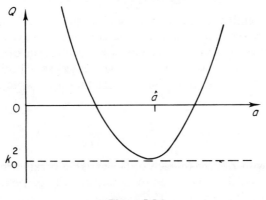

Figure 5.5.1

with the variable parameter a. Clearly, the optimum value of the parameter is $\hat{a}$. Whilst a large number of parameters in the trial function does increase its flexibility, the separate determination of the whole set $\{a\}$ adds considerably to the labour, and in practice some compromise between accuracy and expediency is generally chosen.

From the preceding discussion it might seem that the variation method only applies to the lowest or ground eigenstate, but in fact the excited states can be similarly investigated subject now to the trial functions being orthogonal to the lower state eigenfunctions.

Our initial choice of trial function is made on an intuitive, and consequently *ad hoc* basis. We choose a trial function which we believe will bear some resemblance to the actual distribution. Since we are dealing with strongly perturbed systems we might have grounds for believing that the true distribution has characteristics of two or three known distributions. For example, we know that the ground (unexcited) state of a liquid droplet of radius a_0 in the absence of a gravitational field is spherically symmetric and described by the $P_0^0(\cos \theta)$ harmonic. In the presence of a strong gravitational field we might suppose that $[P_1^0(\cos \theta)]^2$ contributions will develop so as to distort the droplet. The very important *linear combination representation* leads us to the trial function

$$\psi(\mathbf{r}) = a_0[P_0^{0\,2}(\cos \theta) + a_1 P_1^{0\,2}(\cos \theta)] \tag{5.5.8}$$

for the description of the distorted surface, where the variational parameters a_0, a_1 would have to be determined so as to minimize the free energy Q. Clearly, the situation $a_1 = 0$ (spherical droplet) would correspond to minimum surface energy but high gravitational potential energy, and vice versa as a_0 decreases and a_1 increases. There will generally exist some optimum choice of a_0 and a_1 which will yield a minimum value of Q. The linear combination representation is so important because if we write a general trial function

$$\tilde{\psi} = a_0\psi + a_1\psi_1 + \cdots + a_n\psi_n \tag{5.5.9}$$

in terms of some members of the complete set (cf. equation (5.5.2)), then equation (5.5.9) represents a *hybrid distribution* whose components have been chosen on the grounds of physical intuition. We now ask ourselves the question what are the values of the parameters $a_0, a_1, \ldots, a_n$ which yield the optimum form for the trial function? The variation principle tells that if we write down the conditions

$$\frac{\partial Q}{\partial a_0} = 0, \quad \frac{\partial Q}{\partial a_1} = 0, \ldots, \frac{\partial Q}{\partial a_n} = 0 \tag{5.5.10}$$

i.e. minimize Q with respect to each of the variable parameters, we obtain n *secular equations* which may be solved simultaneously to yield the set of coefficients $\{a\}$. Thus, from equation (5.5.4) we have

$$Q = \frac{\int_a^b \tilde{\psi}\mathscr{L}\tilde{\psi}\,\mathrm{d}x}{\int_a^b h\tilde{\psi}\tilde{\psi}\,\mathrm{d}x} \tag{5.5.11}$$

and if, for example, we discuss a two-component trial function $\tilde{\psi} = a_1\tilde{\psi}_1 + a_2\tilde{\psi}_2$ (cf. equation (5.5.8)), then substitution in the expression for Q yields:

$$Q = \frac{a_1^2 \int \psi_1 \mathscr{L}\psi_1 \,\mathrm{d}x + 2a_1 a_2 \int \psi_1 \mathscr{L}\psi_2 \,\mathrm{d}x + a_2^2 \int \psi_2 \mathscr{L}\psi_2 \,\mathrm{d}x}{a_1^2 \int h\psi_1\psi_1 \,\mathrm{d}x + 2a_1 a_2 \int h\psi_1\psi_2 \,\mathrm{d}x + a_2^2 \int h\psi_2\psi_2 \,\mathrm{d}x} \tag{5.5.12}$$

where the Hermitian property of the operator $\mathscr{L}$ enables us to write $\int \psi_1 \mathscr{L} \psi_2^* \, dx = (\int \psi_2^* \mathscr{L} \psi_1 \, dx)^*$: * denotes the complex conjugate. Equation (5.5.12) may be conveniently rewritten

$$Q = \frac{a_1^2 \mathscr{L}_{11} + 2a_1 a_2 \mathscr{L}_{12} + a_2^2 \mathscr{L}_{22}}{a_1^2 S_{11} + 2a_1 a_2 S_{12} + a_2^2 S_{22}} \tag{5.5.13}$$

where

$$\mathscr{L}_{mn} = \int_a^b \psi_m \mathscr{L} \psi_n \, dx \, ; \quad S_{mn} = \int_a^b h\psi_m \psi_n \, dx$$

Variationally minimizing Q with respect to a_1, a_2, (i.e. taking $(\partial Q/\partial a_1)_{a_2} = (\partial Q/\partial a_2)_{a_1} = 0$) we obtain the secular equations

$$a_1(\mathscr{L}_{11} - QS_{11}) + a_2(\mathscr{L}_{12} - QS_{12}) = 0$$
$$a_1(\mathscr{L}_{12} - QS_{12}) + a_2(\mathscr{L}_{22} - QS_{22}) = 0 \tag{5.5.14}$$

The condition that these equations should have a non-trivial solution for a_1, a_2 is that the *secular determinant* of coefficients should vanish:

$$\begin{vmatrix} (\mathscr{L}_{11} - QS_{11}) & (\mathscr{L}_{12} - QS_{12}) \\ (\mathscr{L}_{12} - QS_{12}) & (\mathscr{L}_{22} - QS_{22}) \end{vmatrix} = 0 \tag{5.5.15}$$

Evaluation of this determinant will yield a quadratic for Q, and since all the $\mathscr{L}_{mn}$ and S_{mn} are known it can be solved for Q. For a linear trial function containing n components, equation (5.5.15) will be a polynomial of nth degree in Q, and there will, of course, be n roots, of which the smallest is the one we require. The higher roots correspond to excited states, which they describe somewhat less accurately than the ground state. Having determined the lowest value of Q we may now go back to equation (5.5.14) and finish the determination of a_1 and a_2, or more precisely, the ratio a_1/a_2.

Evaluation of the determinant (5.5.15) illustrates a very important point which generally arises in hybrid situations—the so-called *non-crossing rule*. If the functions ψ_1 and ψ_2 are separately normalized so that $S_{11} = S_{22} = 1$, and $(\mathscr{L}_{11}/S_{11}) = Q_1(\mathscr{L}_{22}/S_{22}) = Q_2$, then if we designate the determinant $\det(Q)$:

$$\det(Q) = (Q - Q_1)(Q - Q_2) - (\mathscr{L}_{12} - QS_{12})^2 \tag{5.5.16}$$

it will have roots at A and B in Figure 5.5.2, corresponding to the first two eigenvalues. The important point is that the roots A, B always lie above and below the Q_1, Q_2 *separately* identified with the distributions ψ_1 and ψ_2. This means that if Q_1 and Q_2 depend upon some parameter ξ, then even if the $Q_1(\xi)$ and $Q_2(\xi)$ curves cross, the two final curves associated with the roots A, B *do not*. This is shown in Figure 5.5.3. This will generally apply regardless of the number of roots. We have already seen some evidence of this in Section 2.9.

As a simple example of the application of the variation method, suppose we consider the fundamental mode of vibration of a stretched string. We know, in

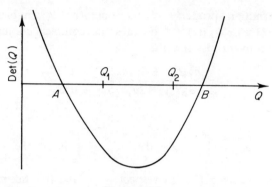

Figure 5.5.2

fact, that the lowest eigenvalue k_0^2 is $\pi^2 T/a^2\rho$, and the corresponding eigenfunction is $\psi = A\sin(\pi x/a)$ for a string of length a, tension T and density ρ. Nonetheless, suppose we propose the parabolic trial function $\tilde{\psi} = x^2 - ax$. This satisfies the boundary conditions that $\tilde{\psi}(0, a) = 0$. Writing the wave equation in Sturm–Liouville form we identify the operator

$$\mathcal{L} = -T\frac{d^2}{dx^2} \tag{5.5.17}$$

whereupon, from equation (5.5.11) we obtain the lowest eigenvalue as

$$\tilde{k}_0^2 = 10T/a^2\rho \tag{5.5.18}$$

which we see corresponds quite closely (to within $1\frac{1}{2}\%$) to the exact value of $\pi^2 T/a^2\rho = 9{\cdot}87T/a^2\rho$. We might carry the calculation further and try the more flexible trial function $\tilde{\psi} = x^n - ax^{n-1}$. In this case we would variationally optimize n by solving $(\partial Q/\partial n) = 0$.

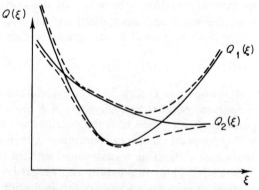

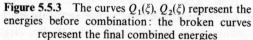

Figure 5.5.3 The curves $Q_1(\xi)$, $Q_2(\xi)$ represent the energies before combination: the broken curves represent the final combined energies

Another example of the application of the variation method is provided by the distribution of electronic charge in the hydrogen atom. Again, we must choose a trial function which satisfies the boundary conditions. Since we require the wavefunction to tend to zero at large distances, the function

$$\tilde{\psi} = e^{-ar} \tag{5.5.19}$$

appears to be satisfactory, where a is a positive constant. Writing the Schrödinger equation in Sturm–Liouville form we find

$$\mathcal{L} = \frac{-\hbar^2}{2m_e} \nabla^2 - \frac{e^2}{r} \tag{5.5.20}$$

which, in this case, is the system Hamiltonian, H.

Since

$$\nabla^2 \tilde{\psi} = \left(a^2 - \frac{2a}{r} \right) e^{-ar}$$

we have, since the weighting function h is unity for the Schrödinger equation,

$$E = \frac{\int_0^\infty \tilde{\psi} H \tilde{\psi} \, d\tau}{\int_0^\infty \tilde{\psi} \tilde{\psi} \, d\tau} = \frac{a^2 \hbar^2}{2m_e} - ae^2 \tag{5.5.21}$$

The optimum value of the parameter a, i.e. which minimizes the energy E, is given by $(\partial E/\partial a) = 0$, whereupon

$$a = \frac{m_e e^2}{\hbar^2} \tag{5.5.22}$$

That is, the optimum form of our trial function is

$$\tilde{\psi}_{opt} = e^{-m_e e^2 r/\hbar^2} \tag{5.5.23}$$

We have been particularly fortunate in our initial choice of function, for comparison of equation (5.5.23) with the first entry in Table 4.7.1 shows that our optimized trial function $\tilde{\psi}_{opt}$ is, in fact, the *exact* result. Substitution of this result for a (equation (5.5.22)) in equation (5.5.21) yields the ground state energy of the hydrogen atom as

$$E_{opt} = -\frac{m_e e^4}{2\hbar^2} \tag{5.5.24}$$

which again is the exact value. Our initial choice of $\tilde{\psi}$ might have been less fortunate—we could have proposed that $\tilde{\psi}$ had the more flexible form e^{-ar^n}. The optimum value of the paramters a and n could then be determined from the secular equations $(\partial E/\partial a) = (\partial E/\partial n) = 0$, in which case we would have found that E was a minimum when $n = 1 \cdot 0$ and $a = m_e e^2/\hbar^2$, as before.

5.6 Quantum Molecular Physics: Molecular Orbital Theory

A particularly interesting application of the variational approximation developed in the last section concerns the determination of the eigenfunctions and eigenvalues of simple molecular structures. Moreover the equilibrium bond lengths and energies are determined in the calculation together with energies of dissociation and ionization. The particular approach we shall use here to estimate the form of the trial function is to take a linear combination of atomic orbitals (LCAO) to represent the molecular orbital. Such a linear combination inevitably involves concepts of hybridization, and subsequent refinement of the trial function will involve arguments of an essentially physical kind.

If we consider the hydrogen molecular ion H_2^+ it is quite clear that whilst the single electron is in the vicinity of nucleus A its eigenfunction is reasonably well described by ψ_A, and whilst in the vicinity of B, by ψ_B. It is true that there will always be some interaction between the electron and both nuclei, but we may incorporate this modification later. Thus, we propose the trial function

$$\tilde{\psi} = \psi_A \pm \psi_B \tag{5.6.1}$$

this representing the simplest linear combination of hydrogenic orbitals. It is, incidentally, perfectly possible to write down the exact bicentric Hamiltonian for the H_2^+ system, incorporating both electron–nuclear interactions, and solve the resulting Schrödinger equation in prolate spheroidal coordinates (Section 4.14). The solution is extremely tedious and difficult, and the physical implications nowhere near as clear as in the LCAO approach we are about to present.

We shall take this opportunity to calculate the energy associated with the trial function (5.6.1). Thus, writing

$$E = \frac{\int \tilde{\psi} H \tilde{\psi}\, d\tau}{\int \tilde{\psi}\tilde{\psi}\, d\tau}$$

we have

$$
\begin{aligned}
&= \frac{\int (\psi_A \pm \psi_B) H(\psi_A \pm \psi_B)\, d\tau}{\int (\psi_A \pm \psi_B)^2\, d\tau} \\
&= \frac{E_A + E_B \pm 2\int \psi_A H \psi_B\, d\tau}{\int \psi_A^2\, d\tau + \int \psi_B^2\, d\tau \pm 2\int \psi_A \psi_B\, d\tau}
\end{aligned}
\tag{5.6.2}
$$

If the atomic orbitals ψ_A, ψ_B are separately normalized so that $\int \psi_A^2\, d\tau = 1.0$, etc., we have

$$E = \frac{E_A \pm \beta}{1 \pm S} \tag{5.6.3}$$

since $\psi_A = \psi_B$, $E_A = E_B$. β is termed the *resonance integral* and represents a measure of the energy associated with the electronic coupling between the systems A and B: $\int \psi_A H \psi_B\, d\tau$. S is termed the *overlap integral*, is necessarily positive, and clearly vanishes if the atomic centres are not sufficiently close, as does β (Figure 5.6.1).

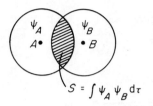

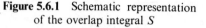

Figure 5.6.1 Schematic representation of the overlap integral S

S may be zero if the two wavefunctions ψ_A, ψ_B are of the wrong symmetry type to be combined in which case the wavefunctions are orthogonal, and cannot combine to form a molecular orbital. Taking the $+(-)$ sign with the $+(-)$ sign in equation (5.6.3) we see that *provided there is sufficient coupling*, the doubly degenerate level E_A splits under the action of the perturbation to yield *two* non-degenerate levels *both* of which are associated with the entire H_2^+ system.

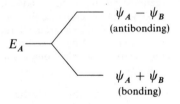

Since β is in fact a negative quantity, the orbitals $(\psi_A - \psi_B)$, $(\psi_A + \psi_B)$ are assymetrically distributed about the degenerate level E_A. $\psi = \psi_A + \psi_B$ is seen to be associated with the lower energy and is therefore termed the *bonding orbital*. The orbital $\psi_A - \psi_B$ is correspondingly the *antibonding orbital*. Quite obviously, the closer the two atomic centres are brought together the greater will be the resonance integral β, and the lower energy level will become lower and lower, and the upper, higher and higher. There will also be the Coulomb repulsion acting between the two screened nuclear charges, and this will act to disrupt the system. The bonding and antibonding energy curves as a function of internuclear separation R_{AB} are shown in Figure 5.6.2 together with the Coulomb term. The total energies in the two cases are shown by the full curves. It is seen that only the bonding orbital $\psi_A + \psi_B$ yields a stable minimum: the antibonding potential curve represents an unstable structure spontaneously disintegrating with the emission of energy: $H_2^+ \rightarrow p + H$. From Figure 5.6.2 we conclude that the H_2^+ molecule can exist, and has an equilibrium bond length of about 2·5 Bohr radii (just over 1 Å). From a knowledge of the form of the potential function we could, if we wished, go on to calculate the vibrational frequency of the system together with the dissociation energy—both experimentally observable.

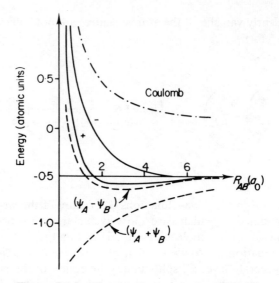

Figure 5.6.2 The energy curves for H_2^+. The 'bonding' curve corresponding to the stable H_2^+ molecule is denoted by the $+$ sign, and is formed from the repulsive internuclear Coulombic term ($-\cdot-\cdot-$) and the $(\psi_A + \psi_B)$ electronic term. The 'antibonding' curve denoted by $-$ is formed from the Coulombic and $(\psi_A - \psi_B)$ electronic component

In Figure 5.6.3(a, b) we show the electronic density distributions in the bonding and antibonding modes. The stability of the bonding distribution is immediately understood in terms of the localization of electronic charge between the nuclei, acting as a sort of electronic 'cement'. Note that the concentration there is greater than the simple sum of contributions from two separate atomic orbitals since for these hydrogenic orbitals $(\psi_A + \psi_B)^2 > \psi_A^2 + \psi_B^2$. Precisely the converse is true for the antibonding orbital $\psi_A - \psi_B$.

What, however, do we write for ψ_A and ψ_B in the trial function $\tilde{\psi} = \psi_A \pm \psi_B$? A first guess would be the ground state atomic orbitals of hydrogen:

$$\psi_A = \left(\frac{1}{\pi a_0^3}\right)^{\frac{1}{2}} e^{-r_a/r_0}; \quad \psi_B = \left(\frac{1}{\pi a_0^3}\right)^{\frac{1}{2}} e^{-r_b/r_0} \tag{5.6.4}$$

An immediate improvement would be to incorporate the effect of an *effective* nuclear charge of $+ce$, where c is to be variationally determined. In this case the trial function becomes

$$\tilde{\psi} = \psi_A \pm \psi_B$$
$$= \left(\frac{c^3}{\pi a_0^3}\right)^{\frac{1}{2}} e^{-cr_a/r_0} \pm \left(\frac{c^3}{\pi a_0^3}\right)^{\frac{1}{2}} e^{-cr_b/a_0} \tag{5.6.5}$$

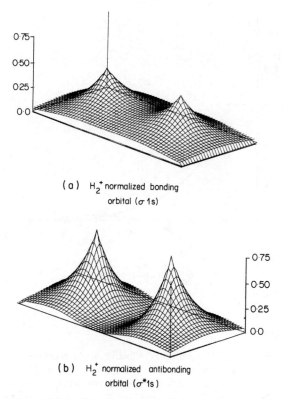

(a) H_2^+ normalized bonding
orbital (σ 1s)

(b) H_2^+ normalized antibonding
orbital (σ^*1s)

Figure 5.6.3 The distribution of electronic charge in the bonding (a) and antibonding (b) orbitals of the H_2^+ molecule. We see that the bonding orbital corresponds to an enhanced electron density along the internuclear axis thereby stabilizing the molecule

The optimum value of c will, of course, depend upon the internuclear separation, and at large values of R_{AB} the orbitals will degenerate into their hydrogenic form with $c = 1\cdot0$. This is shown in Figure 5.6.4 for the bonding and antibonding cases. As we see, in the antibonding case the orbitals retain a distorted hydrogenic form and the screening factor c is relatively insensitive to the internuclear separation. The bonding mode, however, shows a very sensitive dependence upon the bond length. At the observed internuclear distance of $R = 2\cdot0a_0$, the effective nuclear charge for the bonding orbitals is $ce = 1\cdot24e$.

Of course, we have so far entirely neglected the deviations from sphericity of ψ_A and ψ_B as the two atomic orbitals approach each other. We should make some concession to polarity of the atomic orbitals. We may incorporate an angular dependence in the atomic orbital as follows,

$$\psi_A = \left(\frac{c^3}{\pi a_0^3}\right)^{\frac{1}{2}} e^{-cr_a/a_0}\{1 + \lambda P_1^0(\cos\theta)\}$$

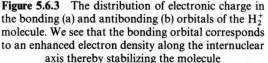

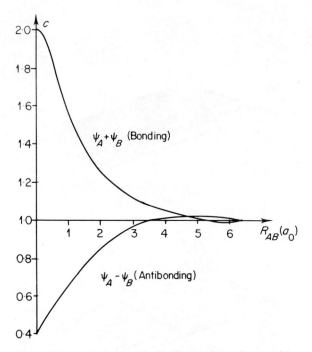

Figure 5.6.4 Variation of effective nuclear charge with internuclear separation for the bonding $(\psi_A + \psi_B)$ and antibonding $(\psi_A - \psi_B)$ orbitals

where both λ and c are to be variationally optimized in the trial function $\tilde{\psi} = \psi_A \pm \psi_B$. The resulting potential curve is shown in Figure 5.6.5(c), and the result is indistinguishable from the exact curve.

In the case of the homonuclear diatomics such as H_2, we have to account additionally for the electron–electron repulsion between the two ground state electrons of opposite spin. A general word concerning the *aufbau* principle is appropriate here, since it enables us to determine the electronic configuration of a molecule, the orbitals having been determined. According to the *aufbau-prinzip* electrons are fed into the vacant orbitals with due regard to the Pauli principle concerning opposed spins in the same eigenstate, and starting at the lowest energy level and working upwards. The *aufbauprinzip* applied to *atomic* orbitals gives an immediate demonstration of the periodic system of the elements.

We have already discussed the molecular hydrogen ion: the single electron will be located in the bonding orbital as shown in Figure 5.6.6—the arrow schematically indicating the spin. Similarly, in the hydrogen molecule, a second electron may be accommodated in the bonding orbital—but with opposed spin (Figure 5.6.6). We neglect at this stage any spin–orbit interaction: the levels shown in Figure 5.6.6 should in fact show fine structure according as

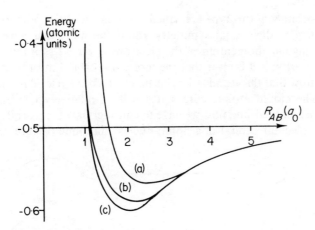

Figure 5.6.5 Various stages of approximation for the energy function of the H_2^+ molecule (a) using hydrogen functions, (b) introduction of the screening constant c in ψ_A and ψ_B, and (c), introduction of polarization terms in ψ_A, ψ_B

the magnetic moment associated with the electron spin is parallel or antiparallel to the orbital magnetic moment. Let us continue applying the *aufbau* principle in the present situation. Suppose we bring together a helium atom and a hydrogen atom so that again we have a bonding orbital ($\psi_{He} + \psi_H$) and an antibonding orbital ($\psi_{He} - \psi_H$). Because the exclusion principle prevents us from placing more than two electrons in one molecular orbital, the three electrons must occupy the two molecular orbitals in the following way $(\psi_{bonding})^2 (\psi_{antibonding})^1$. Thus the molecule HeH: should just be stable by the energy $(2E_b - E_a)$. Such a molecule is known in the vapour phase.

Finally, suppose we bring together two He atoms. Then we have four electrons to be accommodated in two molecular orbitals; two must go in the bonding and two in the antibonding orbitals as shown in Figure 5.6.6. Thus, according to the LCAO approximation the He_2 molecule is unstable and spontaneously disintegrates. A similar argument may be framed for all of the inert gases, and this explains why they are monatomic.

So far we have only considered cases in which the only important molecular orbitals are formed by overlap of an s-orbital on each of the two atoms.

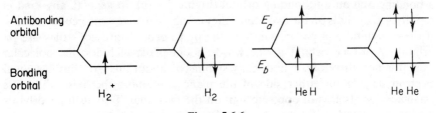

Figure 5.6.6

Molecular orbitals of this type are called σ orbitals and the property which so classifies them is their axial symmetry about the molecular axis. Moreover the orbital angular momentum of the electrons is zero. From the linear combination of s orbitals it is clear that the molecular orbital must be σ.

Suppose now that the atomic s level is filled whilst the p level is only partially occupied. The closed atomic core states will be relatively unaffected as the molecule is assembled, and now we have to consider how the p orbitals combine to form a molecular orbital in a homo-nuclear diatomic (Figure 5.6.7). There

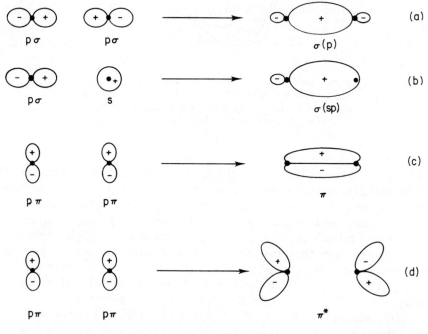

Figure 5.6.7

will, in general, be three orthogonal p orbitals each accommodating two electrons. If we define the internuclear axis as the x-axis, two p_x orbitals may combine to form a bonding $(p_x(A) + p_x(B))$ and an antibonding $(p_x(A) - p_x(B))$ orbital. These are also σ molecular orbitals (Figure 5.6.7a). Note that a p_x orbital on one atom may also combine with an s orbital on the other to produce a bonding and an antibonding orbital (Figure 5.6.7b). In general, any kind of s–p_x hybrid orbitals on the two atoms may combine to give σ molecular orbitals. If the p_x orbital is a $p\sigma$ orbital, then the p_y, p_z orbitals are not—they are $p\pi$ orbitals. A π-type orbital is one whose nodal plane includes the molecular axis. It is not, therefore, cylindrically symmetric about this axis, but has equal electron density on either side of the plane containing the axis, whilst the wavefunction itself is of opposite sign on the two sides. Two such $p\pi$ orbitals can be combined into a bonding πMO, $p\pi(A) + p\pi(B)$, and an antibonding

πMO, $p\pi(A) - p\pi(B)$. These two are simply denoted π and π^* respectively—the asterisk generally indicating the antibonding state. Their formation is shown in Figure 5.6.7(c), (d).

Let us consider an element in the first short period, which has 2s and 2p orbitals in its valence shell. When two such atoms are combined into a homonuclear diatomic molecule, the two sets of atomic orbitals may combine into various MOs. Before we are able to apply the *aufbau* principle to determine the electronic configurations of the diatomic molecule, we must know the relative energies of the molecular orbitals. The results of theoretical and experimental study have shown that the order is generally as shown in Figure 5.6.8. This

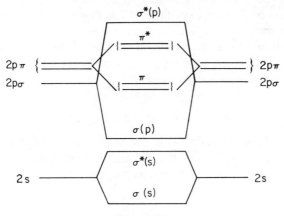

Figure 5.6.8

diagram introduces the concept of *degenerate orbitals*. Notice in particular that the π and π^*MOs each consist of two bracketed levels, which is meant to indicate that they have the same energy. One of the π orbitals is formed from a combination of the p_y orbitals, and the other from the p_z orbitals (see Figure 5.6.7c). These π orbitals are completely equivalent, except that their spatial orientations differ by 90°. Since the diatom is symmetrical about its axis, this difference in orientation cannot make any difference in its energies.

The treatment of heteronuclear diatomics by LCAO theory is not fundamentally different from the treatment of the homonuclear diatomics, except that the molecular orbitals are not symmetric with respect to a plane perpendicular to and bisecting the internuclear axis. The molecular orbitals are still constructed by forming linear combinations of atomic orbitals on the two atoms, but since the atoms are now different we must write $\psi_A \pm \lambda\psi_B$, where λ is not, in general, equal to unity. Thus these molecular orbitals will represent polar bonding, and as examples we may consider HCl and NO.

In discussing HCl we find it necessary to mention explicitly another factor influencing the stability of a bonding MO. Even if two atomic orbitals are capable of combining from the point of view of symmetry, the extent to which they will actually couple will depend on whether their energies are comparable to begin with. If their energies are vastly different, they will scarcely mix at all. We might remind the reader of the very close classical analogue involving the coupling of two linked pendula of different fundamental frequencies.

5.7 Quantum Molecular Physics: Valence Bond Theory

We have seen in Section 5.5 that having assumed a trial function $\tilde{\psi}$, its parameters may be variationally optimized to minimize the associated energy. The trial function may be progressively refined on a physical basis to account for the various aspects which we consider should be represented in a molecular orbital. When the bond energy closely approaches the experimental value we may have some confidence that the trial function then describes the essential features of the molecular orbital. In the molecular orbital approach we established the MOs as a linear combination of atomic orbitals, and then fed in the electrons according to an *aufbauprinzip*, taking due regard of Pauli exclusion and pairing the electrons accordingly.

An alternative approach is to try the trial function

$$\tilde{\psi} = \psi_A(1)\psi_B(2) \tag{5.7.1}$$

for the orbital of the molecule AB. A and B are the constituent atoms, and the implication of equation (5.7.1) is that electron 1 is permanently associated with atom A, and 2 with B. Heitler and London proposed the immediate improvement which allows *exchange* of the electrons 1 and 2 between the centres A and B as follows

$$\tilde{\psi} = \psi_A(1)\psi_B(2) \pm \psi_A(2)\psi_B(1) \tag{5.7.2}$$

In the LCAO approximation discussed in Section 5.6 the orbitals are intrinsically polycentric: not until we take the step proposed by Heitler and London in equation (5.7.2) is the essentially polycentric nature of the orbitals accounted for. For a homonuclear diatomic molecule such as H_2, $\psi_A = \psi_B$ and both components in equation (5.7.2) appear with equal weight—a variational determination on the trial function would also show this, but it is evident from the symmetry. Again we have bonding and antibonding orbitals, and insertion of equation (5.7.2) in

$$E = \frac{\int \tilde{\psi} H \tilde{\psi} \, d\tau}{\int \tilde{\psi} \tilde{\psi} \, d\tau}$$

yields

$$E = 2E_A + \frac{Q \pm J}{1 \pm S^2} \tag{5.7.3}$$

where S is again the overlap integral and Q is the *Coulomb integral*:

$$Q = \iint \psi_A(1)\psi_B(2)(H - 2E_A)\psi_A(1)\psi_B(2) \, d\tau_1 \, d\tau_2 \tag{5.7.4}$$

and represents the total electrostatic energy associated with the repulsion between the two nuclei, the repulsion between the two charge clouds, and the cross-attraction between the charge cloud on B and nucleus A, and *vice versa*. This term is quite classical. The term J is the *exchange integral*

$$J = \iint \psi_A(1)\psi_B(2)(H - 2E_A)\psi_A(2)\psi_B(1)\, d\tau_1\, d\tau_2 \qquad (5.7.5)$$

which should be compared carefully with Q. This term is entirely non-classical, and represents the energy associated with the delocalization and exchange of the electron from its parent nucleus.

Just as in the LCAO approach, we have to decide what we are to use for the functions ψ_A, ψ_B. In the case of the hydrogen molecule we may try the function

$$\psi_A = \left(\frac{c^3}{\pi a_0^3}\right)^{\frac{1}{2}} e^{-cr_0/a_0} \qquad (5.7.6)$$

and variationally determine the screening constant c. For the bonding orbital the Wang function (5.7.6) shows a qualitatively similar dependence upon the internuclear distance as shown in Figure 5.6.4. Next we may account for deviations from sphericity of the component atomic orbitals, and write

$$\psi_A = \left(\frac{c^3}{\pi a_0^3}\right)^{\frac{1}{2}} e^{-cr_a/a_0}(1 + \lambda P_1^0(\cos\theta)) \qquad (5.7.7)$$

where λ is now to be variationally optimized, precisely as in the LCAO approach.

Now, we have not accounted for the possibility of finding both electrons temporarily in the vicinity of one or other of the atomic centres: the bonding trial function $\psi_A(1)\psi_B(2) + \psi_A(2)\psi_B(1)$ is entirely symmetrical. If both electrons are on A, then they would have a wavefunction $\psi_A(1)\psi_A(2)$, where $\psi_A \sim e^{-c'r_a/a_0}$. We may therefore introduce an 'ionic' wavefunction

$$\tilde{\psi}_{\text{ion}} = \psi_A(1)\psi_A(2) + \psi_B(1)\psi_B(2) \qquad (5.7.8)$$

and write the total trial function as

$$\tilde{\psi} = \tilde{\psi}_{\text{cov}} + \alpha\tilde{\psi}_{\text{ion}} \qquad (5.7.9)$$

where the initial trial function, before the introduction of ionic components, is termed the *covalent* function. α must now be variationally determined, and turns out to be 0·25 at the observed equilibrium separation. Clearly the covalent aspect of the bonding dominates. The gradual improvement of the energy curve as the trial function is refined is shown in Figure 5.7.1. We could continue to increase the flexibility of the trial function, and indeed, Kolos and Roothaan, using a fifty-term function for the ground state of H_2 obtain a maximum binding energy of 4·7467 eV compared with the experimental value of 4·7466 $\pm$ 0·0007 eV, and an equilibrium separation of 0·741 Å, identical to the experimental value.

228

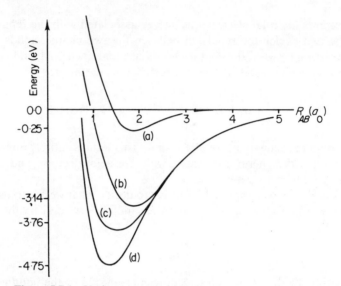

Figure 5.7.1 Various stages of approximation for the energy
function of the H_2 molecule

    (a) $\bar{\psi} = \psi_A(1)\psi_B(2)$
    (b) $\bar{\psi} = \psi_A(1)\psi_B(2) + \psi_A(2)\psi_B(1)$
    (c) $\bar{\psi}_{(b)}$ + Wang screening
    (d) Kolos and Roothaan 50-term function; experiment

5.8 Hybridization

In the previous two sections we have seen how the two principal approximate
theories of bonding in conjunction with the variation principle lead to a
successful theory of molecular structure and bond lengths and energies. We
restricted the discussion of homo- and heteronuclear diatomics where over-
lapping of s or p atomic bonding orbitals led to a stable diatom. There is no
reason at all why we should not go on to consider a polyatomic system such as
H_2O. The ground state electronic configuration of the oxygen atom is
$(1s)^2(2s)^2(2p_x)^2(2p_y)^1(2p_z)^1$ which suggests that the $(2p_y)(2p_z)$ orbitals are
available for bonding. The other orbitals are not, of course, since they are
already doubly occupied by electrons with opposed spins. The ground state
hydrogen orbitals are the singly occupied (1s) functions, and since the $2p_x, 2p_y,$
$2p_z$ orbitals are mutually orthogonal, we might anticipate the HOH bond angle
in water to be 90°. Since H–H repulsions can open the bond angle out to 95°
and the experimental value is 104·5°, it follows that the binding cannot be pure p.

Much more dramatic is the tetragonal carbon coordination observed in an
enormous number of organic molecules, not to mention the tetragonal pure
carbon structure of diamond. The electronic configuration of carbon in its
ground state is $(1s)^2(2s)^2(2p_x)^1(2p_y)^1$ which immediately suggests directed
valence with bond angles of 90° between the bonding p orbitals.

It is found, however, that if we raise the carbon into the excited configuration $(1s)^2(2s)^1(2p_x)^1(2p_y)^1(2p_z)^1$ and then allow the $(2s)^1(2p_x)^1(2p_y)^1(2p_z)^1$ electrons to *hybridize*, we obtain a *lower* final bond energy which more than offsets that required to promote a 2s electron to the $2p_z$ state. It was Slater and Pauling separately who pointed out that if the 2s and 2p orbitals were hybridized with each other, then a much higher electron density can be produced between the C and X atoms in the C—X bond, so that the bond is much stronger.

From the s, p_x, p_y, p_z orbitals it is possible to form the hybrids

$$\psi_1 = \alpha_1 s + \beta_1 p_x + \gamma_1 p_y + \delta_1 p_z$$
$$\psi_2 = \alpha_2 s + \beta_2 p_x + \gamma_2 p_y + \delta_2 p_z$$
$$\psi_3 = \alpha_3 s + \beta_3 p_x + \gamma_3 p_y + \delta_3 p_z \tag{5.8.1}$$
$$\psi_4 = \alpha_4 s + \beta_4 p_x + \gamma_4 p_y + \delta_4 p_z$$

where the coefficients $\alpha_i, \beta_i, \gamma_i, \delta_i$, must be selected subject to orthogonality and normalization of the functions. A particularly interesting set of tetrahedral bonds is formed as follows:

$$_{(111)}\psi_1 = \tfrac{1}{2}(s + p_x + p_y + p_z)$$
$$_{(1\bar{1}\bar{1})}\psi_2 = \tfrac{1}{2}(s + p_x - p_y - p_z)$$
$$_{(\bar{1}1\bar{1})}\psi_3 = \tfrac{1}{2}(s - p_x + p_y - p_z) \tag{5.8.2}$$
$$_{(\bar{1}11)}\psi_4 = \tfrac{1}{2}(s - p_x - p_y + p_z)$$

where the subscripts (111), etc., represent the directions of the orbitals within a unit cube. Each bond is entirely equivalent to the others and represents the tetrahedral coordination shown for the methane molecule CH_4 shown in Figure 5.8.1. The valence bonds established between the tetrahedral carbon

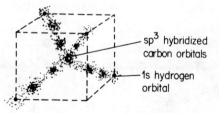

sp³ hybridized carbon orbitals

1s hydrogen orbital

Figure 5.8.1 Schematic coordination of the tetrahedral methane (CH_4) molecule

orbitals and the s orbitals of the hydrogens are doubly occupied as schematically indicated in Figure 5.8.1. The postulate of the tetrahedral carbon coordination in classical stereochemistry does not necessarily require the coordination to be regular with a bond angle of 109° 28′. The values of the coefficients $\alpha_i, \ldots, \delta_i$ in equation (5.8.1) need not all be equal, although they will need to satisfy the

230

orthogonality and normalization conditions. We may now readily understand
the structure of the ethane molecule

It has been shown by Kemp and Pitzer that the potential energy of the two
methyl groups fluctuates rapidly as they are rotated about the C—C bond. The
minimum energy configuration occurs for the staggered arrangement, whilst
the maximum occurs in the eclipsed orientation both shown in Figure 5.8.2.

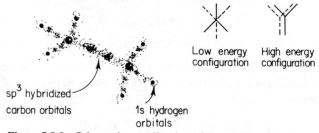

Figure 5.8.2   Schematic coordination of the ethane molecule

This is readily understood in terms of the interaction of the two methyl groups;
the potential energy of the molecule changes by about 3 kcal/mole, the potential
function having three maxima and minima corresponding to the trigonal
symmetry of the methyl groups. The quantized vibrational modes of the
methyl molecules are discussed in terms of the solution to Mathieu's equation
in Section 2.9.

As we know, the electrical properties of semiconductors may be drastically
altered by the addition of either of two types of impurity. The two types are
designated n-type donor and p-type acceptor, and are illustrated in Figure
5.8.3 for a germanium crystal which has a tetragonal structure arising from sp^3

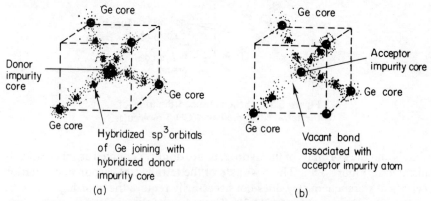

Figure 5.8.3   Schematic effect of the substitutional inclusion of a *donor* impurity core
(a), and (b) an *acceptor* impurity core in Ge

hybridization in the promoted configuration $(3d)^{10}(4s)^1(4p_x)^1(4p_y)^1(4p_z)^1$. (Ground state configuration of Ge: $(3d)^{10}(4s)^2(4p_x)^1(4p_y)^1$). In Figure 5.8.3(a) we show a unit cube containing an n-type or *donor* impurity atom at its centre. The donor atom normally has *five* valence electrons. When it substitutes for a Ge atom in the crystal, four of its five valence electrons share in the covalent bonds, and are relatively tightly bound. The fifth valence electron can occupy a less tightly bound state in which its wavefunction is essentially localized around the donor atom. This 'extra' electron, however, can also receive energy from thermal vibrations, and be excited into the conduction band where it may participate in conduction (Figure 5.8.4a). Some pentavalent atoms which are

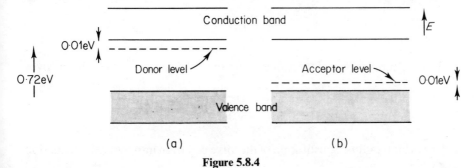

Figure 5.8.4

often used substitutionally as donor impurities in Ge and Si are P, As and Sb. In Ge the energies of the fifth valence electrons of the impurities lie about 0·01 eV below the bottom of the conduction band. This small energy difference should be contrasted with the band-gap energy which is about 0·72 eV for Ge. The energy of the bound state of the fifth valence electron is called the donor energy or the donor level.

The excess electron moves in the Coulomb potential $e/\varepsilon r$ of the impurity ion, where ε is the dielectric constant of the medium. The factor $1/\varepsilon$ takes account of the reduction in the Coulomb force between charges caused by the electric polarization of the medium. This treatment is valid for orbits large in comparison with the distance between atoms, and for slow motions of the electron such that the time required to pass an atom is long in comparison with the period of the motion of the core electrons, conditions satisfied by the valence electron.

The binding energy of the donor impurity may now be very simply calculated. The Bohr theory of the hydrogen atom may be readily modified to take into account both the dielectric constant of the medium, and the effective mass of an electron in the periodic potential of the crystal. Formally, e^2 must be replaced by e^2/ε and m by m^* in the standard results.

Consider, then, the Bohr model of the hydrogen atom using the old quantum theory. The quantization condition is

$$\oint p \, dq = nh$$

where n is the principal quantum number, giving

$$m^*r^2\omega(2\pi) = nh$$

The total energy is the sum of the kinetic and potential energy

$$E = -\frac{e^2}{\varepsilon r} + \frac{1}{2}m^*r^2\omega^2$$

and the force equation is

$$m^*\omega^2 r = \frac{e^2}{\varepsilon r^2}$$

and so the substitution of the value of ω from the quantization condition gives

$$r = \frac{\varepsilon n^2 \hbar^2}{e^2 m^*}$$

for the orbital radius. Substitution of the force equation into the energy equation gives

$$E = -\frac{e^2}{2\varepsilon r}$$

and eliminating r with the help of the above equation yields

$$E = -\frac{e^4 m^*}{2\varepsilon^2 n^2 \hbar^2}$$

The application of these formulae to Ge and Si is complicated by the anisotropic effective mass of the conduction electrons in these crystals. However, ε enters as the square, whereas m^* enters as the first power only. We may obtain a general impression of the impurity levels by using an average value of the anisotropic effective masses. The ionization potential of hydrogen is 13·6 eV, using $\varepsilon = 1$, $m^* = m$ and $n = 1$. For Ge the ionization potential on the present model should be reduced with respect to hydrogen by the factor $m^*/m\varepsilon^2$, giving 0·0065 eV, whilst the corresponding result for Si is 0·025 eV.

Just as an electron may be bound to a pentavalent impurity, a hole may be bound to a trivalent impurity in Ge or Si. The acceptor problem is similar in principle to that for electrons, and the ionization energies are not unlike those for donors. The equivalent Bohr model carries over in principle for holes just as for electrons, except that in Ge and Si there is an orbital degeneracy at the top of the valence band which enormously complicates the theoretical treatment.

Problems

1 'Electrons have a spin $\frac{1}{2}$ and a magnetic moment of 1 Bohr magneton'. Explain what is meant by this statement, and outline the evidence for it.

In calculating energy levels for a helium atom, one allows for electrostatic repulsion between the two electrons but disregards the magnetic interaction between their spins. What is the justification for this procedure?

2 A one-dimensional oscillator of charge e is described by a wave equation

$$-\frac{\hbar^2}{2m}\frac{d^2}{dx^2}u_n(x) + \tfrac{1}{2}kx^2u_n(x) = E_n u_n(x)$$

with solutions $u_n(x)$ associated with energies $E_n = (n + \frac{1}{2})\hbar\omega$, where $\omega^2 = k/m$. A weak electric field of strength ε is then applied in the x-direction. By considering the perturbation expansion for the new ground-state energy

$$E'_0 = E_0 + \int_{-\infty}^{\infty} u_0^* U u_0 \, dx + \sum_{n=1}^{\infty} \frac{|\int_{-\infty}^{\infty} u_n^* U u_0 \, dx|^2}{E_0 - E_n}$$

where U is the perturbing change in the potential energy, show that there is no change in energy of the ground-state to first order in the electric field, and calculate the second-order change.

Show that the problem can be solved exactly by a suitable shift of origin, and compare the resulting ground-state energy with the value obtained from the perturbation theory formula. (You may assume the property of oscillator eigenfunctions,

$$\int_{-\infty}^{\infty} u_n^* x u_m \, dx = \left(\frac{n+1}{2}\right)^{\frac{1}{2}} \alpha^{-1}, \quad m = n+1$$

$$\left(\frac{n}{2}\right)^{\frac{1}{2}} \alpha^{-1}, \quad m = n-1$$

$$0, \quad \text{otherwise}$$

where $\alpha = (m^k/\hbar^2)^{\frac{1}{4}}$.)

The Fourier Transform

Whilst we have used the power series representation of an arbitrary function $f(x)$ at many points in this book, such expansions often fail completely due either to the functional form of $f(x)$, or otherwise converge over limited regions of the variable x. For example, the function $1/(1 - x)$ may be expanded as

$$\frac{1}{1 - x} = 1 + x + x^2 + x^3 + \cdots \tag{I.1}$$

but the radius of convergence is $|x| = 1$.

A more powerful series representation is the expansion of an arbitrary function $f(x)$ in terms of an orthonormal set in general, and the Fourier series of sines and cosines in particular. In this way all but the most extraordinary functions may be Fourier-expanded, and the range of convergence is always over the interval $-\pi < x < \pi$. Thus we may write

$$f(x) = \sum_{n=0}^{\infty} \{a_n \sin nx + b_n \cos nx\} \tag{I.2}$$

where the coefficients a_n, b_n remain to be determined. The physical significance of such an expansion may be readily understood as follows. Suppose we have a function $f(x)$ of period 2π (Figure I.1a). This particular function is synthesized from a combination of the trigonometric functions $-\sin x$ and $+\sin^2 4x$. The amplitudes a_1 and a_2 of these two components may be conveniently displayed as a frequency spectrum, Figure I.1(b). Thus, a periodic but complicated oscillation may be generally represented as a complex synthesis of sinusoidal

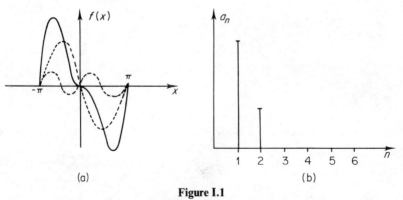

(a)   (b)

Figure I.1

functions, and the frequency distribution gives us the distribution of amplitude amongst the frequencies.

We may save ourselves some trouble in evaluating the set of coefficients a_n, b_n by noting whether the function $f(x)$ is *odd* or *even* about $x = 0$. Clearly, in attempting to represent an even function $f(x)$ we should only wish to retain the even trigonometric functions, $\cos nx$: the coefficients a_n being set to zero. The converse is true for odd functions. This may be very easily demonstrated, and is left to the reader to verify.

Provided the function $f(x)$ *is* periodic, there is no difficulty in scaling the variable x to make it of period 2π. Thus, if we write

$$f(x) = \sum_{n=0}^{\infty} \left\{ a_n \sin\left(\frac{n\pi x}{L}\right) + b_n \cos\left(\frac{n\pi x}{L}\right) \right\} \tag{I.3}$$

then the expansion is valid in the range $-L < x < L$.

We may exploit the orthogonality of the trigonometric functions in equation (I.3) to determine the Fourier coefficients a_n, b_n. Thus, multiplying equation (I.3) throughout by $\sin(m\pi x/L)$ we obtain

$$a_m = \frac{1}{L} \int_{-L}^{L} f(x) \sin\left(\frac{m\pi x}{L}\right) dx \tag{I.4}$$

since

$$\int_{-L}^{L} \sin\left(\frac{m\pi x}{L}\right) \cos\left(\frac{n\pi x}{L}\right) dx = 0$$

$$\left. \begin{array}{l} \displaystyle\int_{-L}^{L} \sin\left(\frac{m\pi x}{L}\right) \sin\left(\frac{n\pi x}{L}\right) dx \\[2ex] \displaystyle\int_{-L}^{L} \cos\left(\frac{m\pi x}{L}\right) \cos\left(\frac{n\pi x}{L}\right) dx \end{array} \right\} = 0 \qquad m \neq n$$

$$\int_{-L}^{L} \sin^2\left(\frac{m\pi x}{L}\right) dx = \int_{-L}^{L} \cos^2\left(\frac{m\pi x}{L}\right) dx = L.$$

We similarly determine

$$b_m = \frac{1}{L} \int_{-L}^{L} f(x) \cos\left(\frac{m\pi x}{L}\right) dx \tag{I.5}$$

Of course, this orthogonality property applies over specified limits of the integration only (See Section 1.7).

If we attempt to represent some function, say a saw-tooth wave, in terms of a partial sum of Fourier components, then we see that the approximate wave form becomes a better and better representation as the number of terms is

236

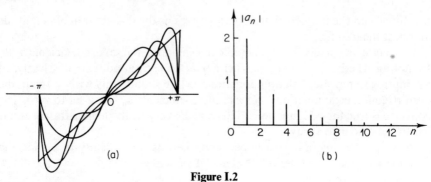

Figure I.2

increased; successive approximations are shown in Figure I.2(a). In this case, over the interval $-\pi < x < \pi$ we may represent $f(x)$ as x. The function is odd, and the coefficients a_m are given by equation (I.4) as

$$\alpha_m = \frac{1}{\pi} \int_{-\pi}^{\pi} x \sin mx \, dx = (-1)^{m-1}(2/m) \tag{I.6}$$

The series is, then,

$$x = 2 \sin x - \sin 2x + \tfrac{2}{3} \sin 3x - \tfrac{1}{2} \sin 4x + \tfrac{2}{5} \sin 5x - \cdots \tag{I.7}$$

and the associated frequency spectrum is shown in Figure I.2(b). The function is *never* exactly reproduced, even with an infinite number of terms. This, known as the *Gibbs phenomenon*, occurs because the Fourier series for a function $f(x)$ does not converge uniformly in the neighbourhood of a discontinuity in $f(x)$. Similar effects occur for expansions in terms of any orthonormal set.

It is often convenient to write the Fourier series in complex form:

$$f(x) = \sum_{n=-\infty}^{\infty} A_n e^{in\pi x/L} \tag{I.8}$$

where

$$A_n = \frac{1}{2L} \int_{-L}^{L} f(x) e^{-in\pi x/L} \, dx \tag{I.9}$$

This brings us to the *Fourier transform* of a non-periodic function, or rather a function whose period is infinite. In other words, we let $L \to \infty$. The function $f(x)$ is then represented over the entire range of the variable $-\infty < x < \infty$. First of all we observe that the sum (I.8) may be converted into an integral since the increments $\Delta n/L$ in the series become infinitely small. Writing,

$$k = n\pi/L, \quad \Delta k = \Delta n(\pi/L) = \pi/L$$

since $\Delta n = 1$, k becomes a continuous variable. From equation (I.8)

$$f(x) = \frac{L}{\pi} \int_{-\infty}^{\infty} A_n e^{ikx} \, dx \tag{I.10}$$

A_n is now a continuous variable of k, and so from equation (I.9)

$$f(k) = LA_n = \tfrac{1}{2} \int_{-\infty}^{\infty} f(x)\,e^{-ikx}\,dx \qquad (I.11)$$

Writing equations (I.10) and (I.11) in slightly more symmetrical form, we have the Fourier transforms

$$f(k) = \frac{1}{\sqrt{(2\pi)}} \int_{-\infty}^{\infty} f(x)\,e^{-ikx}\,dx \qquad (I.12)$$

$$f(x) = \frac{1}{\sqrt{(2\pi)}} \int_{-\infty}^{\infty} f(k)\,e^{ikx}\,dk \qquad (I.13)$$

Extensive tables of Fourier transforms are available in the literature, and we will not develop any of these here. Nevertheless, we may make the general observation that if $f(x)$ is the real space function then $f(k)$ represents the associated frequency spectrum.

It should be realized that $f(k)$ is effectively the *continuous form* of the discrete spectrum (Figure I.2b) arising for a periodic function.

If, for example, we investigate the lattice dynamics of a three-dimensional system of coupled oscillators, such as a crystalline lattice we may determine a quantity termed the *velocity autocorrelation*. This is defined as

$$\psi(t) = \frac{\langle \mathbf{v}_i(t) \cdot \mathbf{v}_i(0) \rangle}{\langle \mathbf{v}_i(0)^2 \rangle} \qquad (I.14)$$

where $\mathbf{v}_i(0)$ and $\mathbf{v}_i(t)$ represent the velocity vectors of the ith atom at times 0 and t respectively. The angular brackets $\langle \ \rangle$ represent the taking of an ensemble average. For a solid, $\psi(t)$ is shown schematically in Figure I.3. The form of the function reflects the generally vibratory characteristic of the lattice, but nevertheless exhibits the *decorrelation* of the vector $\mathbf{v}_i(t)$ with respect to its initial magnitude

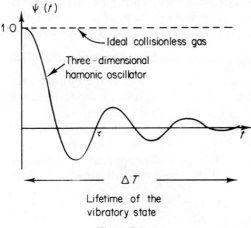

Figure I.3

238

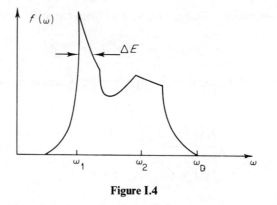

Figure I.4

and direction $v_i(0)$. The time τ approximately represents the period of lattice vibration. If we Fourier transform $\psi(t)$, we obtain the frequency spectrum $\psi(\omega)$ shown in Figure I.4. The pronounced peaks at ω_1 and ω_2 reflect the principal modes of lattice vibration. The distribution cuts off quite sharply in the region of ω_D, the Debye frequency, since the lattice cannot sustain vibrations of indefinitely high frequency. Again, very low frequency vibratory modes do not appear: these are associated with diffusive motions which are largely absent in a crystalline lattice. The area beneath the distribution $\psi(\omega)$ is proportional to the number of modes of vibration, and is generally normalized to be equal to this quantity.

We note that the Heisenberg uncertainty principal may be approached in terms of the general properties of waves. Clearly, the more localized a wave packet is in real space the greater the frequency range required to synthesize it. The two limiting cases are represented by the sine wave extending to $\pm\infty$, having a delta function frequency spectrum on one hand, and the delta function wave packet which has a constant frequency distribution. If we regard the dimension of the wave packet as representing the spatial uncertainty in the location of the associated particle, and the width of the k-space distribution as the uncertainty in momentum, then quite clearly we have a relationship of the form

$$\Delta x \Delta p \sim \text{constant}$$

There are many other such examples of the Fourier transform, perhaps the most familiar being that of diffraction of a monochromatic plane wave by a three-dimensional lattice. The lattice effectively acts as a harmonic analyser—integral relationships between the crystallographic periodicities and that of the incident radiation (Bragg's Law) producing a reciprocal image or diffraction pattern of the lattice. This particular application of Fourier transformation is so adequately discussed elsewhere we shall consider it no further here.

The Laplace Transform

Whilst the essentials of many physical problems may be expressed in terms of a linear differential equation together with the appropriate boundary conditions, its solution is not always an easy task. The Laplace transform method of solution is a particularly powerful approach, and whilst we shall not be concerned here with its mathematical foundation, we shall try to indicate its use as an analytical tool in the solution of physical problems.

The Laplace transform of a function $f(x)$ is defined as

$$\tilde{f}(p) = \int_0^\infty f(x) e^{-px} \, dx \tag{II.1}$$

where it is assumed that the integral exists. We may readily evaluate these transforms for simple functions $f(x)$. For example, if $f(x) = 1$, then

$$\tilde{f}(p) = \int_0^\infty e^{-px} \, dx = \frac{1}{p} \tag{II.2}$$

of if $f(x) = e^{ax} \sin bx$, then

$$\tilde{f}(p) = \int_0^\infty e^{(a-p)x} \sin bx \, dx = \frac{b}{(p-a)^2 + b^2} \tag{II.3}$$

and so on. We may build up a 'dictionary' of such transforms for various functional forms $f(x)$, and these are widely tabulated in the literature, a few examples being given in Table II.1.

Laplace transforms of the derivatives df/dx, $d^2f/dx^2, \ldots, d^nf/dx^n, \ldots$ may also be determined. Thus, if we require the transform of df/dx,

$$\tilde{f}(p) = \int_0^\infty e^{-px} \frac{df}{dx} \, dx$$

$$= [e^{-px}f]_0^\infty + p \int_0^\infty e^{-px} f \, dx$$

$$= -f(0) + p\tilde{f}(p) \tag{II.4}$$

where $f(0)$ represents the value of $f(x)$ at $x = 0$. This is, of course, provided

$$\underset{x \to \infty}{\text{Lt}} \ (e^{-px}f) = 0$$

We may go on to show that the transform of d^2f/dx^2 is

$$\tilde{f}(p) = -f^{(1)}(0) - pf(0) + p^2\tilde{f}(p) \tag{II.5}$$

239

240

Table II.1. Laplace transforms (a, b, c are constants)

$f(x)$	$\bar{f}(p)$
$f'(x)$	$p\bar{f}(p) - f(0)$
$f^{(n)}(x)$	$p^n\bar{f}(p) - p^{n-1}f(0) - p^{n-2}f'(0) - \cdots - f^{(n-1)}(0)$
1	$1/p$
$\delta(x - x_0)$	e^{-px_0}
x	$1/p^2$
x^n	$n!/p^{n+1}$
e^{ax}	$1/(p - a)$
$x^n e^{ax}$	$n!(p - a)^{n+1}$
$\sin ax$	$a/(p^2 + a^2)$
$\cos ax$	$p/(p^2 + a^2)$
$e^{bx} \sin ax$	$a/\{(p - b)^2 + a^2\}$
$e^{bx} \cos ax$	$(p - b)/\{(p - b)^2 + a^2\}$
$\dfrac{e^{-at} - e^{-bt}}{b - a}$	$\dfrac{1}{(p + a)(p + b)}$
$\dfrac{a e^{-at} - b e^{-bt}}{a - b}$	$\dfrac{p}{(p + a)(p + b)}$
$\dfrac{(b - c)e^{-at} + (c - a)e^{-bt} + (a - b)e^{-ct}}{(a - b)(b - c)(c - a)}$	$\dfrac{1}{(p + a)(p + b)(p + c)}$
$\text{erfc}\,\dfrac{a}{2\sqrt{t}}$	$\dfrac{1}{p}\exp{-a\sqrt{p}}$

where $f(0)$ and $f^{(1)}(0)$ represent the evaluation of $f(x)$, df/dx at $x = 0$. It may be quite generally shown that

$$\int_0^\infty e^{-px}\frac{d^nf}{dx^n}\,dx = -(f^{(n-1)}(0) + pf^{(n-2)}(0) + \cdots + p^{n-1}f(0)) + p^n\bar{f}(p) \quad \text{(II.6)}$$

provided the integral converges.

The solution of a linear differential equation by the method of Laplace transformation is then reduced to the following mechanical stages:

(i) Form the Laplace transform of the differential equation. This is known as the *subsidiary equation*.

(ii) Manipulate the subsidiary equation into a form in which it may be readily solved for $\bar{f}(p)$.

(iii) Solve for $\bar{f}(p)$ and use the table of transforms 'inversely' to determine $f(x)$, the solution of the original differential equation.

An example will clarify the method.

Suppose we wish to solve

$$\ddot{\psi} + k^2\psi = a \cos nt \qquad (II.7)$$

and $\psi = \dot{\psi} = 0$ at $t = 0$. The subsidiary equation is, from the table of transforms,

$$(p^2 + k^2)\tilde{\psi}(p) = \frac{ap}{p^2 + n^2}$$

i.e.

$$\tilde{\psi}(p) = \frac{ap}{(p^2 + n^2)(p^2 + k^2)} \qquad (II.8)$$

Hence,

$$\psi(t) = \frac{a}{k^2 - n^2}(\cos nt - \cos kt) \qquad (II.9)$$

Notice how some manipulation of the expression (II.8) for $\tilde{\psi}(p)$ is required before it is in a recognizable form for inversion. That the inverse is a *unique* solution is by no means obvious: the proof of this uniqueness is given in Lerch's theorem. We shall not consider this theorem any further here.

The *convolution theorem* enables us to construct the inverse of the product of two transforms whose separate inverses are known. Thus, if by the operator $\mathscr{L}$ we mean taking the Laplace transform

$$\tilde{f}(p) = \mathscr{L}\{f(x)\}$$

and if by $\mathscr{L}^{-1}$ we mean taking the inverse

$$f(x) = \mathscr{L}^{-1}\{\tilde{f}(p)\}$$

then the convolution theorem states

$$\mathscr{L}^{-1}\{\tilde{f}(p)\tilde{g}(p)\} = \int_0^x f(\xi)g(x - \xi)\,d\xi \qquad (II.10)$$

where $\mathscr{L}^{-1}\{\tilde{f}(p)\} = f(x)$, $\mathscr{L}^{-1}\{\tilde{g}(p)\} = g(x)$. Thus, since

$$\frac{1}{p - a} = \mathscr{L}\{e^{ax}\}$$

$$\frac{1}{p - b} = \mathscr{L}\{e^{bx}\}$$

by the convolution theorem we have

$$\frac{1}{(p - a)(p - b)} = \mathscr{L}\left\{\int_0^x e^{a\xi} e^{b(x - \xi)}\,d\xi\right\}$$

$$= \mathscr{L}\left\{e^{bx}\left(\frac{1}{a - b}e^{(a - b)\xi}\right)_0^x\right\} \qquad (II.11)$$

$$= \mathscr{L}\left\{\frac{1}{a - b}(e^{ax} - e^{bx})\right\} \qquad (II.12)$$

APPENDIX III

The Bessel Function

Introduction

The Bessel differential equation has the form

$$\frac{d^2R}{dr^2} + \frac{1}{r}\frac{dR}{dr} + \left(k^2 - \frac{n^2}{r^2}\right)R = 0 \tag{III.1}$$

where n is a real positive number, integral or fractional. The condition that $R(r)$ is single-valued requires n to be integral: we shall, nevertheless, be concerned with both integral and non-integral orders of Bessel function.

On changing to the variable $x = kr$, equation (III.1) takes the form

$$\frac{d^2R}{dx^2} + \frac{1}{x}\frac{dR}{dx} + \left(1 - \frac{n^2}{x^2}\right)R = 0 \tag{III.2}$$

Equation (III.2) will generally have two linearly independent forms of solution $J_n(x)$ and $Y_n(x)$, these being the Bessel functions of the first and second kinds respectively, and of order n. k, moreover, is real.

Two other forms of linearly independent solution arise, $I_n(x)$ and $K_n(x)$, being the hyperbolic Bessel functions of the first and second kinds, again of order n, but this time of *imaginary argument*. That is, k being imaginary.

There are Bessel functions of the third kind, known as Hankel functions, given as

$$H_n^{(1)}(x) = J_n(x) + iY_n(x)$$

$$H_n^{(2)}(x) = J_n(x) - iY_n(x)$$

We shall not discuss these functions further here.

Solution of Bessel's Equation

Equation (III.2) has a *regular singular point* at $r = 0$. In attempting series solutions of such equations we generally assume a power-series expansion of the form

$$R(x) = x^m \sum_{j=0}^{\infty} a_j x^j \tag{III.3}$$

The reason for this choice will become apparent when we come to determine the coefficients a_j. If from equation (III.3) we determine R', R'' and substitute in equation (III.2) we find, on comparing coefficients of terms of the same order of x,

$$a_{j-2} = -a_j[(m + j + n)(m + j - n)] \tag{III.4}$$

242

This *recurrence (or recursion) relation* relates alternate coefficients in the series expansion. Such a recurrence relationship must always be established before the power series (III.3) can be expanded. The recurrence is not always between alternate coefficients, of course—it may be between adjacent or more widely separated coefficients.

If we set $j = 0$, we obtain from equation (III.4)

$$a_{-2} = -a_0(m + n)(m - n)$$

Since the coefficient of r^{-2} is zero (otherwise the series (III.3) would become infinite at the origin) we obtain the *indicial equation*

$$0 = -a_0(m + n)(m - n) \tag{III.5}$$

from which we conclude, if $a_0 \neq 0$ (which it must not be if $a_2, a_4, \ldots, a_{2j}$ are to be non-zero from the recurrence relation),

$$m = \pm n \tag{III.6}$$

If, in our initial expression of the series (III.3) we set $m = 0$, since $n \neq 0$ we should, in equation (III.5), have to conclude that $a_0, \ldots, a_{2j}$ were all zero.

Taking $m = +n$ in the recurrence relationship we immediately obtain

$$a_{j-2} = -a_j j(2n + j)$$

giving

$$R(x) = J_n(x) = \frac{1}{n!}\left(\frac{x}{2}\right)^n \left\{ 1 - \frac{1}{n+1}\left(\frac{x}{2}\right)^2 + \frac{1}{(n+1)(n+2)}\frac{1}{2!}\left(\frac{x}{2}\right)^4 - \cdots \right\}$$

$$= \sum_{j=0}^{\infty} \frac{(-1)^j}{j(n+j)!}\left(\frac{x}{2}\right)^{n+2j} \tag{III.7}$$

Unfortunately, for integral values of n a second linearly independent solution to Bessel's equation is not provided by the function $J_{-n}(x)$. Indeed, it may be quite simply shown that $J_{-n}(x) = (-1)^n J_n(x)$. The second independent solution in the integral case must be found separately, and is in fact the Bessel function of the second kind $Y_n(x)$, sometimes known as the Weber, or Neumann function, $N_n(x)$, and is defined as

$$Y_n(x) = \frac{J_n(x)\cos(n\pi) - J_{-n}(x)}{\sin(n\pi)} \tag{III.8}$$

where the limiting value of this expression is taken if n is zero or an integer. The first two sets of independent solutions $[J_0(x), Y_0(x)]$ and $[J_1(x), Y_1(x)]$ are shown in Figure III.1. The functions of the second kind are all infinite at the origin, and clearly cannot be present in any interior solution which requires the function to be finite at the origin. $Y_n(x)$ may, however, appear in exterior solutions.

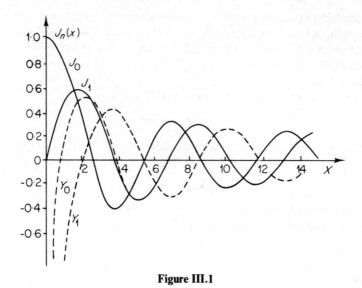

Figure III.1

The preceding analysis in no way depends on n being integral, and we may extend our definition (III.7) to explicitly include non-integral orders of Bessel function in terms of the gamma function, thus:

$$R(x) = \sum_{j=0}^{\infty} \frac{(-1)^j}{j!\Gamma(n + j + 1)} \left(\frac{x}{2}\right)^{n+2j} \tag{III.9}$$

In this case the Bessel functions J_n, J_{-n}, where n is now non-integral, *do* provide linearly independent solutions. Examining, for example, the solutions for $n = +\frac{1}{3}$, the exponents $\frac{1}{3}, \frac{7}{3}, \frac{13}{3}, \ldots$, arise, whilst for $n = -\frac{1}{3}$ we obtain the exponents $-\frac{1}{3}, \frac{5}{3}, \frac{11}{3}, \ldots$. Clearly these are two distinct functions, and the general solution to Bessel's equation for *non-integral n* is $R(x) = AJ_n + BJ_{-n}$. Non-integral Bessel functions of the second kind may of course be defined by equation (III.8) where we no longer have to pass to the limit.

The hyperbolic Bessel functions of the first and second kinds, $I_n(x)$, $K_n(x)$, sometimes known as the Kelvin functions, may be separated into their real and imaginary components. Thus, $I_n(x)$ may be written (from equation (III.7)) for positive integral, or zero, n:

$$I_n(x) = \frac{J_n(ix)}{i^n} = \frac{1}{n!}\left(\frac{x}{2}\right)^n \left\{ 1 + \frac{1}{(n+1)}\left(\frac{x}{2}\right)^2 + \frac{1}{(n+1)(n+2)}\frac{1}{2!}\left(\frac{x}{2}\right)^4 + \cdots \right\} \tag{III.10}$$

so that, for example

$$I_0(x) = \left\{ 1 + \left(\frac{x}{2}\right)^2 + \frac{1}{2\cdot 2!}\left(\frac{x}{2}\right)^4 + \cdots \right\} \tag{III.11}$$

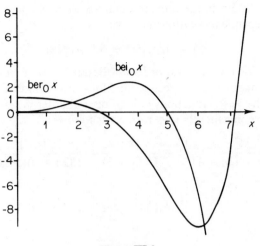

Figure III.2

Similarly,

$$K_n(x) = \frac{\pi i}{2} \cdot i^n H_n^{(1)}(ix)$$

The function $J_n(xi\sqrt{i})$ separates into its real and imaginary components, thus:

$$[\text{ber}_n(x) + i\,\text{bei}_n(x)] = e^{in\pi/2} I_n(x\,e^{i\pi/4}) \tag{III.12}$$

The two functions $\text{ber}_0(x)$, $\text{bei}_0(x)$ are shown in Figure III.2. The hyperbolic functions of the second kind $K_n(x)$ may be similarly expressed:

$$[\text{ker}_n(x) + i\,\text{kei}_n(x)] = e^{-in\pi/2} K_n(x\,e^{i\pi/4})$$

A differential equation which frequently describes the radial solution in systems of spherical symmetry is the following:

$$\frac{d^2R}{dr^2} + \frac{2}{r}\frac{dR}{dr} + \left(k^2 - \frac{l(l+1)}{r^2}\right)R = 0 \tag{III.13}$$

which should be compared with its cylindrical counterpart, equation (III.1).

If we make the transformation $\mathscr{R}(r) = r^{\frac{1}{2}}R(r)$, then equation (III.13) can be expressed in recognizable form:

$$\frac{d^2\mathscr{R}}{dr^2} + \frac{1}{r}\frac{d\mathscr{R}}{dr} + \left(k^2 - \frac{(l+\frac{1}{2})^2}{r^2}\right)\mathscr{R} = 0 \tag{III.14}$$

and now substituting $x = kr$ we obtain Bessel's equation of half-integral order

$$\frac{d^2\mathscr{R}}{dx^2} + \frac{1}{x}\frac{d\mathscr{R}}{dx} + \left(1 - \frac{(l+\frac{1}{2})^2}{x^2}\right)\mathscr{R} = 0 \tag{III.15}$$

(cf. equation (III.2)). From the previous analysis we may conclude that acceptable linearly independent solutions are given as

$$\mathcal{R}(x) = AJ_{l+\frac{1}{2}}(x) + BJ_{-(l+\frac{1}{2})}(x)$$

(Remember that for non-integral orders of Bessel function we do not have to seek recourse to functions of the second kind for a second independent solution.)

If we expand the Bessel functions of half-integral order in series form we find that they may be conveniently re-expressed as follows:

$$j_n^{(x)} = \frac{x^n}{1 \cdot 3 \cdot 5 \cdots (2n + 1)} \left\{ 1 - \frac{\frac{1}{2}x^2}{1!(2n + 3)} + \frac{(\frac{1}{2}x^2)^2}{2!(2n + 3)(2n + 5)} - \cdots \right\}$$

$$y_n^{(x)} = \frac{1 \cdot 3 \cdot 5 \cdots (2n - 1)}{x^{n+1}} \left\{ 1 - \frac{\frac{1}{2}x^2}{1!(1 - 2n)} + \frac{(\frac{1}{2}x^2)^2}{2!(1 - 2n)(3 - 2n)} - \cdots \right\}$$

(III.16)

i.e.

$$j_n(x) = \sqrt{(\tfrac{1}{2}\pi/x)}J_{n+\frac{1}{2}}(x) \Big\}$$
$$y_n(x) = \sqrt{(\tfrac{1}{2}\pi/x)}Y_{n+\frac{1}{2}}(x) \Big\}$$

(III.17)

where $j_n(x)$, $y_n(x)$ are the *spherical Bessel functions of the first and second kinds*, respectively. From the expansions (III.16) it is quite easy to show that

$$j_0(x) = \sin x/x, \qquad j_1(x) = \sin x/x^2 - \cos x/x$$

$$y_0(x) = -\cos x/x, \qquad y_1(x) = -\cos x/x^2 - \sin x/x$$

We may continue to define the spherical Hankel functions, or *spherical Bessel functions of the third kind*:

$$h_n^{(1)}(x) = j_n(x) + iy_n(x)$$
$$h_n^{(2)}(x) = j_n(x) - iy_n(x)$$

(III.18)

The pairs $(j_n(x), y_n(x))$ are linearly independent solutions for every n. However, for these half integral functions we have $j_n(x)$ and $j_{-n}(x)$ as linearly independent solutions, and indeed

$$y_n(x) = (-1)^n j_{-n}(x)$$

(III.19)

The functions $(j_0(x), y_0(x))$ and $(j_1(x), y_1(x))$ are shown in Figure III.3.

Properties

The Bessel functions have certain similarities with the simple trigonometric functions $\cos(x)$ (cf. $J_0(x)$) and $\sin(x)$ (cf. $J_1(x)$) in that they have many zeros and properties analogous to $d(\sin x)/dx = \cos x$.

From the expansion (III.7), regrouping immediately leads to the result

$$J_{n-1}(x) + J_{n+1}(x) = \frac{2n}{x} J_n(x)$$

(III.20)

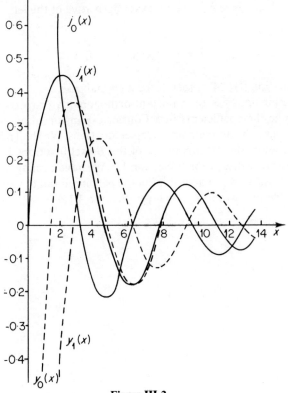

Figure III.3

similarly,

$$J_{n-1}(x) - J_{n+1}(x) = 2J_n'(x) \tag{III.21}$$

Adding these two relations we immediately obtain

$$J_{n-1}(x) = \frac{n}{x}J_n(x) + J_n'(x) \tag{III.22}$$

whilst subtraction yields

$$J_{n+1}(x) = \frac{n}{x}J_n(x) - J_n'(x) \tag{III.23}$$

From equation (III.21) we obtain the important result

$$J_0'(x) = J_1(x) \tag{III.24}$$

since $J_{-1}(x) = -J_1(x)$. A variety of such relations may be obtained in this way, but it is not appropriate to go into them here. An important integral relation will, however, be derived. Multiplication of equation (III.22) by x^n yields

$$x^n J_{n-1}(x) = nx^{n-1}J_n(x) + x^n J_n'(x) \tag{III.25}$$

The right-hand side is seen to be an exact derivative of the left, and so we may write

$$\int x^n J_{n-1}(x)\, dx = x^n J_n(x) \tag{III.26}$$

for which equation (III.24) is seen to be a special case.

Bessel functions possess an important orthogonality property which enables us to determine the coefficients in a Fourier expansion of an arbitrary radial function in terms of a complete set of Bessel functions. We shall not demonstrate the simple proof of the orthogonality of the Bessel functions here—it follows quite easily from the discussion of Section 1.7. We have already shown the use of the Fourier–Bessel series in determining the perturbed radial vibrational modes in the Earth due to a Mohorovičic layer (Section 4.12).

APPENDIX IV

The Hermite Polynomials

Introduction

The Hermite differential equation has the form

$$\frac{d^2H}{dx^2} - 2x\frac{dH}{dx} + (\varepsilon - 1)H = 0 \tag{IV.1}$$

and we attempt a series solution of the form

$$H(x) = \sum_{j=0}^{\infty} a_j x^j \tag{IV.2}$$

If we determine the expressions for H'' and H' from equation (IV.2) and substitute in equation (IV.1), comparison of coefficients in x yields

$$a_j = \begin{cases} \dfrac{(1 - \varepsilon)(5 - \varepsilon)\cdots(2j - 3 - \varepsilon)}{j!}a_0 & \text{for } j \text{ even} \\[3mm] \dfrac{(3 - \varepsilon)(7 - \varepsilon)\cdots(2j - 3 - \varepsilon)}{j!}a_1 & \text{for } j \text{ odd} \end{cases} \tag{IV.3}$$

which means that we may formally write the solution of equation (IV.1) as

$$H(x) = a_0\left\{ \sum_{j\,\text{even}} \frac{(1 - \varepsilon)(5 - \varepsilon)\cdots(2j - 3 - \varepsilon)}{j!}x^j \right\}$$
$$+ a_1\left\{ \sum_{j\,\text{odd}} \frac{(3 - \varepsilon)(7 - \varepsilon)\cdots(2j - 3 - \varepsilon)}{j!}x^j \right\} \tag{IV.4}$$

where each series contains terms which increase by x^2. From equation (IV.3) the ratio of successive coefficients is given as

$$\frac{a_{j+2}}{a_j} = \frac{2j + 1 - \varepsilon}{(j + 1)(j + 2)} \tag{IV.5}$$

which, for large values of j

$$\frac{a_{j+2}}{a_j} \to \frac{2}{j}$$

We now make a direct comparison with the series for e^{x^2}:

$$e^{x^2} = 1 + x^2 + \frac{x^4}{2!} + \cdots + \frac{x^{2n}}{n!} + \frac{x^{2n+2}}{(n + 1)!} + \cdots$$

This series also has its coefficients in the ratio $2/j$ in the limit of large j, and we conclude that both series behave like e^{x^2} as x tends to infinity. The general solution (IV.4) is therefore unacceptable since it diverges at infinity. There are, however, some *special* solutions for which the infinite series happens to terminate at a certain term, the solutions in this case being the *Hermite polynomials*. Thus, if any one of the coefficients a_j happens to be zero, then from the recursion formulae (IV.3) we see that all subsequent coefficients $a_{j+2}, a_{j+4}, \ldots$, are also zero. This will occur for certain values of the constant ε. Thus, in equation (IV.5) we see that the series will terminate as a polynomial at a_j if

$$2j + 1 - \varepsilon = 0$$

i.e.

$$\varepsilon = 2j + 1 \tag{IV.6}$$

whilst the other series is set to zero throughout by taking a_0 or a_1 to be zero. A solution which remains bounded as $x \to \pm\infty$ is therefore only possible if $\varepsilon = 2j + 1$ with integral n. This, in fact, is a quantization condition: in the case of the linear harmonic oscillator (Section 2.8) for example, the energy E is effectively quantized through equation (IV.6) as (equation (2.8.11))

$$E = \frac{h\nu}{2}(2j + 1)$$

If by $H_j(x)$ we mean the termination of the Hermite polynomial at the jth coefficient, then we readily obtain the first few polynomials from equation (IV.5) as

$$H_0(x) = 1$$
$$H_1(x) = 2x$$
$$H_2(x) = 4x^2 - 2$$
$$H_3(x) = 8x^3 - 12x$$
$$H_4(x) = 16x^4 - 48x^2 + 12$$
$$H_5(x) = 32x^5 - 160x^3 + 120x$$

These are shown in Figure IV.1.

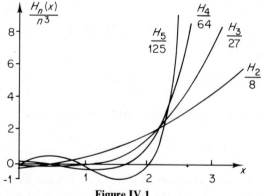

Figure IV.1

On introducing the condition (IV.6) into equation (IV.1) we obtain

$$\frac{d^2 H_j}{dx^2} - 2x\frac{dH_j}{dx} + 2jH_j = 0 \tag{IV.7}$$

and it may be easily shown by direct substitution that this equation has as a solution

$$H_j = K e^{x^2} \frac{d^j}{dx^j} e^{-x^2} \tag{IV.8}$$

where K is an arbitrary constant conventionally chosen as $(-1)^j$.

The principal application of the Hermite functions in this book arises in the solution of the Schrödinger equation for the linear harmonic oscillator. The final form of the eigenfunctions was shown to be (equation (2.8.15))

$$\psi_j(x) = H_j(\sqrt{[\beta x]}) e^{-\beta x^2/2}; \quad \beta = \sqrt{(\hbar/m\omega)} \tag{IV.9}$$

where β is a constant. The factor $e^{-\beta x^2/2}$ arising from the large $|x|$ form of the solution ensures that the function is well-behaved as $x \to \pm\infty$. Whilst the Hermite functions do indeed diverge, they do not diverge faster than $e^{\beta x^2/2}$, and so the above distribution is bounded at infinity.

A few important results will now be stated without proof, as follows.

If may be quite easily shown from the development above that the Hermite polynomials satisfy the recursion relationship

$$\frac{dH_j(x)}{dx} = 2jH_{j-1}(x) \tag{IV.10}$$

and

$$H_{j+1}(x) = 2xH_j(x) - 2jH_{j-1}(x)$$

The important question of normalization of the wavefunction (IV.9) may now be discussed.

As we have seen, the functions H_j are unbounded and cannot therefore be normalized. However, it may be shown that for the wavefunction (IV.9)

$$\int_{-\infty}^{\infty} [H_j(x) e^{-x^2/2}]^2 \, dx = \sqrt{\pi} 2^j j! \tag{IV.11}$$

whereupon we may immediately evaluate the normalization integral:

$$\int_{-\infty}^{\infty} \psi_j(x)\psi_j(x) \, dx = \sqrt{(\hbar/m\omega)} \sqrt{\pi} 2^j j! \tag{IV.12}$$

Hence the normalized wavefunction $\psi_j(x)$ is

$$\psi_j(x) = \left(\frac{m\omega}{\hbar}\right)^{\frac{1}{4}} \frac{1}{(\sqrt{\pi} 2^j j!)^{\frac{1}{2}}} H_j(\sqrt{(m\omega/\hbar)}x) e^{-(m\omega/2\hbar)x^2} \tag{IV.13}$$

These wavefunctions possess an *orthogonality property* according to which

$$\int_{-\infty}^{\infty} \psi_n \psi_m \, dx = \delta_{nm} \qquad \text{(IV.14)}$$

where δ_{nm} is the Kronecker delta symbol ($\delta_{nm} = 0$, $n \neq m$; $\delta_{nm} = 1$, $n = m$). We may therefore expand an arbitrary distribution in terms of a complete orthonormal Fourier–Hermite set:

$$f(x) = \sum_{n=0}^{\infty} a_n \psi_n(x) \qquad \text{(IV.15)}$$

The coefficients a_n may be determined in the usual way by exploiting the orthogonality relationship (IV.14). (See Section 1.7.)

APPENDIX V

The Mathieu Equation

Introduction

The second-order differential equation

$$\frac{d^2\psi}{dx^2} + \beta(x)\psi = 0 \qquad (V.1)$$

where $\beta(x)$ is an even periodic function of x, occurs frequently in physical problems. If $\beta(x)$ is of period π equation (V.1) is known as *Hill's equation* and was first studied by Hill in connection with the moon's orbital motion. This same equation is extensively used in the theory of metallic conduction—Schrödinger's equation cast into Hill's form (V.1) then describing the propagation of an electron in a one-dimensional periodic lattice potential. We shall return to this point in relation to the use of the *Bloch*, or more generally, the *Floquet theorem* which is central to the present discussion. Hill's equation also arises in the theory of particle orbits in the alternating gradient synchrotron. Here the applied magnetic field is the periodic function.

We may expand $\beta(x)$ as a Fourier series of the even (cosine) functions, and retain only the first term, thus

$$\frac{d^2\psi}{dx^2} + (\alpha + \beta \cos 2x)\psi = 0 \qquad (V.2)$$

where α and β are constants. This is *Mathieu's equation*, and finds extensive application in the radial solution of elliptic distributions.

Obviously we are interested in periodic solutions, that is solutions for which

$$\psi(x + 2\pi) = \psi(x) \qquad (V.3a)$$

or in some physical cases, of period π:

$$\psi(x + \pi) = \psi(x) \qquad (V.3b)$$

The theory of Floquet is concerned with the task of finding periodic solutions to equation (V.1) generally, and equation (V.2) in particular, which have period 2π. In fact it turns out that the determination of such solutions is far from trivial, and occur only for certain values of the coefficients α and β in Mathieu's equation. Nevertheless, we may observe at the outset that for the special case $\beta = 0$, equation (V.2) becomes

$$\frac{d^2\psi}{dx^2} + \alpha\psi = 0 \qquad (V.4)$$

253

254

and periodic solutions with period 2π, and which are either *odd* or *even* about $x = 0$, exist as follows:

$$\left.\begin{array}{cc} \psi(x) = 1 & \alpha = 0 \\[2mm] \psi(x) = \begin{pmatrix} \sin x \\ \cos x \end{pmatrix} & \alpha = 1 \\[4mm] \psi(x) = \begin{pmatrix} \sin 2x \\ \cos 2x \end{pmatrix} & \alpha = 4 \\ \vdots & \vdots \end{array}\right\} \qquad \text{(V.5)}$$

The values of α yielding stable oscillatory solutions of period 2π for arbitrary β are shown in the α–β plot in Figure V.1.

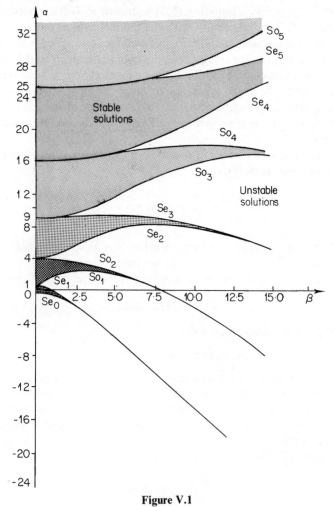

Figure V.1

Floquet was able to show that Mathieu's equation (and indeed Hill's equation) will always have solutions of the form

$$\psi(x) = e^{\mu x} f(x)$$

where $f(x)$ has period 2π. The nature of the solution $\psi(x)$ therefore depends entirely on the nature of μ. μ in fact depends essentially on α and β, and three distinct situations may be recognized.

(i) If μ is zero, or is some integral multiple of i then $\psi(x)$ is of the same period as $f(x)$. For these values of μ we therefore obtain stable oscillatory solutions of period 2π. These are the solutions we require.

(ii) If μ is purely imaginary then the solutions $\psi(x)$ are oscillatory but aperiodic.

(iii) If μ possesses a real part then the solutions are unstable, being non-oscillatory and divergent.

Whilst solutions corresponding to (i) above are required for perfect periodicity (i.e. $e^{\mu x} = 1$) we should nevertheless realize that solutions with complex μ such that $|e^{\mu x}| = 1$ are also physically acceptable since Bloch electronic wavefunctions in a periodic lattice take this form—the charge density $|\psi|^2$ being periodic even though $\psi(x)$ itself is not.

Values of μ not leading to stable oscillatory solutions lead to *attenuated functions*. Mathieu determined the regions of the α–β plane which correspond to stable solutions: the boundaries between the stable and unstable regions correspond to solutions for which $\mu = 0$. These periodic solutions are the Mathieu functions:

$$Se_n (\alpha, \beta, x) = \cos (nx) + \sum_{m \neq n} A_m(\alpha, \beta) \cos (mx)$$

$$So_n (\alpha, \beta, x) = \sin (nx) + \sum_{m \neq n} B_m(\alpha, \beta) \sin (mx)$$

(V.6)

where So, Se connote odd and even functions, respectively. We have yet to determine the coefficients A_m, B_m in terms of the parameters α, β and we should note that only odd m terms are taken if n is odd and only even m terms if n is even. We have already seen that as $\beta \to 0$ so we regain the simple trigonometric functions— the periodic solution lines diverge in pairs from the points $\alpha = n^2$, $\beta = 0$ where n is integral, as we have seen in equation (V.5).

The $\alpha(\beta)$ Functions

The locus of stability in the α–β plane may be quite simply determined as follows. Suppose we adopt the general solution

$$\psi = \tfrac{1}{2}A_0 + \sum_{n=1}^{\infty} \{A_n \cos nx + B_n \sin nx\}$$

(V.7)

determine ψ'', and substitute in the Mathieu differential equation. Comparison of coefficients in $\cos nx$ leads to the three-term recurrence relation.

$$\beta A_{j-2} + 2(\alpha - j^2)A_j + \beta A_{j+2} = 0 \qquad (V.8)$$

(A similar recurrence in B_j occurs in comparing coefficients of $\sin nx$.) Rearrangement yields

$$\frac{A_j}{A_{j-2}} = \frac{-\beta}{2(\alpha - j^2) + \beta \dfrac{A_{j+2}}{A_j}} \qquad (V.9)$$

Using this form of expression we may iteratively substitute for the ratio of coefficients in the denominator, yielding a continued fraction, thus:

$$\frac{A_j}{A_{j-2}} = \cfrac{-\beta}{2(\alpha - j^2) - \cfrac{\beta}{2[\alpha - (j+2)^2] + \beta \dfrac{A_{j+4}}{A_{j+2}} \cdots}} \qquad (V.10)$$

We then substitute for (A_{j+4}/A_{j+2}), and so on. The α–β relation is finally obtained by noting from equation (V.8), or directly by equating coefficients, that for $j = 2$

$$\frac{A_2}{A_0} = -\frac{\alpha}{\beta} \qquad (V.11)$$

Since $A_{-2} = 0$. So from equations (V.10) and (V.11) we have

$$\alpha = \cfrac{\beta^2}{2(\alpha - 4) - \cfrac{\beta^2}{2(\alpha - 16) + \beta\left(\dfrac{A_6}{A_4}\right)}} \qquad (V.12)$$

We see that for large negative α the loci attain their quadratic form. To obtain the loci shown in Figure V.1 we have to know the value of α as $\beta \to 0$. These, in fact, are given as $n^2, n = 0, 1, 2, \ldots$ (see equation (V.5)). Suppose, for example, we require the α–β locus which passes through the point $\alpha = 1$, $\beta = 0$. The successive approximations are then given as:

first approximation

$$\alpha \sim \frac{\beta^2}{2(1 - 4)} = -\frac{\beta^2}{6}$$

second approximation

$$\alpha \sim \cfrac{\beta^2}{2\left(-\dfrac{\beta^2}{6} - 4\right) - \cfrac{\beta^2}{2(1 - 16) + \cdots}}$$

$$\sim -\frac{\beta^2}{8} + \frac{3\beta^4}{640} - \cdots \qquad (V.13)$$

and so on. In this way we are able to establish all the loci shown in Figure V.1. Using the recursion relation (V.8) it is now a relatively easy matter to establish the coefficients A_m (or by a similar analysis, B_m) in terms of α and β. We may eliminate the α-dependence by substitution of the appropriate locus expression, equation (V.13). We should note in particular that although we obtain odd and even solutions (So, Se) just as in the simple trigonometric case (sin, cos), we need to specify whether the solution is periodic in π or 2π: the Mathieu equation like any other second-order differential equation has two independent solutions only.

APPENDIX VI

The Legendre Function

Introduction

Since the majority of problems in spherical polar coordinates differ only in the form of the separated *radial* differential equation, we find a variety of radial solutions—Bessel functions, Laguerre functions, etc.—whilst the angular solutions tend not to vary from problem to problem. Of course, the angular and radial solutions are linked through the separation constant, and this effectively selects which of the complete set of solutions is going to develop.

The conditions that we place on the angular solutions are that they should be everywhere finite and single-valued. Moreover, we require the solution to be continuous, and this, as we shall see, is enough to establish the angular quantization. We may regard the requirement on the angular function to replicate over the interval 0 to 2π as the strict analogue to the perhaps more straightforward quantization imposed by the more usual boundary conditions.

Solution of the Associated Legendre Equation

After separation we generally obtain the second-order differential equation

$$\frac{1}{\sin \theta} \frac{d}{d\theta}\left(\sin \theta \frac{d\Theta}{d\theta}\right) + \left\{\beta - \frac{m^2}{\sin^2 \theta}\right\}\Theta = 0 \qquad (VI.1)$$

m^2 is a constant arising in the separation of the Φ-equation. β is a characteristic or *eigenvalue* of the problem. If we now make the substitution $\mu = \cos \theta$ we obtain

$$\frac{d}{d\theta} = -\sqrt{(1 - \mu^2)}\frac{d}{d\mu}$$

$$\sin \theta = \sqrt{(1 - \mu^2)}$$

$$\Theta(\theta) = P(\mu)$$

and equation (VI.1) becomes

$$\frac{d}{d\mu}\left\{(1 - \mu^2)\frac{dP}{d\mu}\right\} + \left\{\beta - \frac{m^2}{1 - \mu^2}\right\}P = 0 \qquad (VI.2)$$

This differential equation is not yet ready to be solved by the power series method since it contains poles at $\mu = \pm 1$. We therefore require a solution $P(\mu)$ which vanishes at $\mu = \pm 1$. Such a solution is, for $\mu = +1$,

$$P(\mu) = (1 - \mu^2)^\alpha p(\mu) = (1 - \mu^2)^\alpha \sum_{n=0}^{\infty} a_n \mu^n \qquad (VI.3)$$

258

where $p(\mu)$ is the function to be solved by a power series method. We have yet to determine the index α.

Substituting equation (VI.3) in equation (VI.2) we eventually obtain

$$\alpha = \pm \frac{m}{2}$$

We therefore have

$$P(\mu) = (1 - \mu^2)^{|m|/2} \sum_n a_n \mu^n \qquad \text{(VI.4)}$$

We now note that

$$\sum_n a_n \mu^n$$

need not vanish at $\mu = \pm 1$. Having determined the derivatives P'', P' from equation (VI.4) and substituting in equation (VI.2), comparison of coefficients yields the following *recursion relation*

$$a_{n+2} = \frac{(n + m)(n + m + 1) - \beta}{(n + 1)(n + 2)} a_n \qquad \text{(VI.5)}$$

From this relation we may obtain all the even coefficients, given a_0, and all the odd coefficients given a_1. We may therefore write

$$\sum_n a_n \mu^n = a_0 \sum_{\text{even}} + a_1 \sum_{\text{odd}} \qquad \text{(VI.6)}$$

We see that from equation (VI.5), for large n

$$\frac{a_{n+2}}{a_n} \sim 1$$

that is, the series is likely to converge only for $\mu < 1$. What happens, for example, at $\mu = \pm 1$ for large n? The result is indeterminate, as we see

$$P(\mu) = (1 - \mu^2)^{|m|/2} \sum_n a_n \mu^n$$

i.e. at $\mu = \pm 1$

$$P(\mu) = 0 \cdot \infty$$

In fact, a closer examination shows that the 'infiniteness' of the series is greater than the 'zeroness' of $(1 - \mu^2)^{|m|/2} : P(\mu)$ still diverges at $\mu = \pm 1$. The only way we can obtain acceptable solutions that remain finite over the entire interval $-1 \leqslant \mu \leqslant 1$ is for the series to terminate as a polynomial at some term. Thus, if from equation (VI.6) we set

$$(n + m)(n + m + 1) = \beta \qquad \text{(VI.7)}$$

the series breaks off at the nth term becoming a polynomial of degree n. The other series will still diverge of course, but we may dispose of it by setting the initial coefficient a_0 or a_1 to zero, as the case may be.

Since both n and m are positive integers, equation (VI.7) may be written

$$l(l + 1) = \beta \qquad \text{(VI.8)}$$

Condition (VI.8) effectively establishes the quantization of the eigenvalue β, whilst equations (VI.4) and (VI.5) determine the eigenfunctions. The degeneracy associated with the angular solutions is immediately apparent from equation (VI.8), and arises from the various combinations of n and m which give rise to the same l.

If we now re-express equation (VI.4) in the original variable θ, we have

$$P_l^m(\cos \theta) = \sin^{|m|} \theta \sum_{k=0}^{n} a_k \cos^k \theta \qquad \text{(VI.9)}$$

The functions $\Theta(\theta) = P_l^m(\cos \theta)$ are called the *associated Legendre functions*. Moreover, we see from equation (IV.7) that since

$$l = n + m \qquad \text{(VI.10)}$$

and n is a positive or zero integer, then

$$l \geqslant m \qquad \text{(VI.11)}$$

From the solutions $\Theta(\theta)$, $\Phi(\phi)$ we may construct the *surface spherical harmonics* $Y_l^m(\theta, \phi)$. Thus we may write

$$Y_l^m(\theta, \phi) = e^{im\phi} P_l^m(\cos \theta)$$

where the indices are restricted to the values $m \leqslant l$.

For the sake of completeness it should be mentioned that the *associated Legendre functions of the second kind*, $Q_l^m(\mu)$, represent the second solution to the associated Legendre differential equation. These solutions are, in fact, given by the infinite series which was discarded by setting the initial coefficient, either a_0 or a_1, to zero. These solutions become infinite at $\mu = \pm 1$ and are usually rejected on physical grounds. They develop quite naturally, however, in the case of spheroidal coordinate systems.

Properties

The associated spherical harmonics can perhaps be best envisaged from Figure 4.3.1. The first few unassociated (ϕ-independent, i.e. $m = 0$) Legendre functions $P_l(\cos \theta)$ are shown in Figures 4.1.2 and VI.1.

We state without proof *Rodrigues' generating expression* for the unassociated functions

$$P_l(\mu) = \frac{1}{2^l l!} \frac{d}{d\mu^l}(\mu^2 - 1)^l; \quad \mu = \cos \theta \qquad \text{(VI.12)}$$

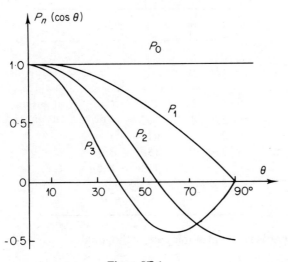

Figure VI.1

That this is so may be seen by direct substitution of equation (VI.12) into the *unassociated* ($m = 0$) differential equation (VI.2). A few of these functions are listed below:

$$P_0(\mu) = 1 \qquad\qquad P_4(\mu) = \tfrac{35}{8}\mu^4 - \tfrac{15}{2}\mu^2 + \tfrac{3}{8}$$

$$P_1(\mu) = \mu \qquad\qquad P_5(\mu) = \tfrac{63}{8}\mu^5 - \tfrac{35}{4}\mu^3 + \tfrac{15}{8}\mu$$

$$P_2(\mu) = \tfrac{3}{2}\mu^2 - \tfrac{1}{2}$$

$$P_3(\mu) = \tfrac{5}{2}\mu^3 - \tfrac{3}{2}\mu^2 \qquad\qquad \mu = \cos\theta$$

The associated Legendre functions $P_l^m(\mu)$ are related to the unassociated functions $P_l(\mu)$ by *Ferrer's formula*:

$$P_l^m(\mu) = (1 - \mu^2)^{|m|/2}\, \frac{d^{|m|}}{d\mu^{|m|}} P_l(\mu) \qquad\qquad (VI.13)$$

or, in conjunction with Rodrigues' formula

$$= \frac{(1 - \mu^2)^{|m|/2}}{2^l l!}\, \frac{d^{l+|m|}}{d\mu^{l+|m|}} (\mu^2 - 1)^l \qquad\qquad (VI.14)$$

That this is so may be shown by direct substitution of equation (VI.14) into the associated differential equation (VI.2). The first few associated functions $P_l^m(\mu)$ are given below:

l	m	$P_l^m(\mu)$
0	0	1
1	0	μ
1	1	$(1 - \mu^2)^{\frac{1}{2}}$
2	0	$\frac{1}{2}(3\mu^2 - 1)$
2	1	$3\mu(1 - \mu^2)^{\frac{1}{2}}$
2	2	$3(1 - \mu^2)$
3	0	$\frac{1}{2}(5\mu^3 - 3\mu)$
3	1	$\frac{3}{2}(1 - \mu^2)^{\frac{1}{2}}(5\mu^2 - 1)$
3	3	$15(1 - \mu^2)\mu$
3	3	$15(1 - \mu^2)^{\frac{3}{2}}$

The surface spherical harmonics are orthogonal:

$$\int_0^{2\pi} \int_0^{\pi} Y_l^m Y_{l'}^{m'} \sin\theta \, d\theta \, d\phi = 0 \quad l \neq l', \quad m \neq m' \tag{VI.15}$$

as are the individual angular functions

$$\left.\begin{array}{l} \int_0^{2\pi} \Phi_m \Phi_{m'} \, d\phi = 0 \quad m \neq m' \\[2mm] \int_0^{\pi} P_l^m P_{l'}^{m'} \sin\theta \, d\theta = 0 \quad \begin{array}{l} l \neq l' \\ m \neq m' \end{array} \end{array}\right\} \tag{VI.16}$$

The normalization factor for the angular functions may be either compounded from the individual normalizations,

$$N_\phi(m) = \left[\int_0^{2\pi} \Phi_m \Phi_m \, d\phi \right]^{-\frac{1}{2}} = \frac{1}{\sqrt{2\pi}}$$

$$N_\theta(l, m) = \left[\int_0^{\pi} P_l^m P_l^m \sin\theta \, d\theta \right]^{-\frac{1}{2}} = \sqrt{\frac{(2l + 1)(l - m)!}{2(l + m)!}}$$

or directly from the surface spherical harmonics

$$N_{\theta,\phi}(l, m) = \left[\int_0^{2\pi} \int_0^{\pi} Y_l^m Y_l^m \sin\theta \, d\theta \, d\phi \right]^{-\frac{1}{2}} = \frac{1}{\sqrt{2\pi}} \sqrt{\frac{(2l + 1)(l - m)!}{2(l + m)!}}$$

Any well-behaved angular distribution of θ and ϕ over the surface of a sphere may be expanded in terms of a spherical harmonic series. This is precisely the angular counterpart of the Fourier expansion of a linear periodic function (see Section 1.7).

APPENDIX VII

The Laguerre Function

Introduction

An important form of the separated radial equation arises in the solution of the hydrogen atom in particular, and the hydrogenic systems in general. Writing the Schrödinger equation for the Coulomb central field problem in spherical polar coordinates we obtain, for the separated radial equation (4.7.5):

$$\frac{d^2\mathscr{R}}{d\rho^2} + \frac{2}{\rho}\frac{d\mathscr{R}}{d\rho} - \left\{\frac{1}{4} - \frac{\lambda}{\rho} + \frac{l(l+1)}{\rho^2}\right\}\mathscr{R} = 0 \qquad \text{(VII.1)}$$

where λ is a constant related to the total energy and the effective nuclear charge of the system. ρ is the reduced radial coordinate. Of particular importance is the constant $l(l+1)$ which has arisen in the separation of the radial and angular equations. It has been shown that for finite, well-behaved angular solutions l is restricted to adopt positive integral values only (Appendix VI: The Legendre function). To this extent the radial and angular solutions are linked.

As usual, we are looking for well behaved, single valued bounded solutions, and we shall find that these conditions are sufficient to establish a quantization of the total energy, just as the restrictions on l established a quantization of the angular momentum in the case of the angular solution of the Schrödinger equation.

We have seen that at large values of ρ (Section 4.7) equation (VII.1) reduces to

$$\frac{d^2\mathscr{R}}{d\rho^2} - \frac{1}{4}\mathscr{R} \sim 0$$

which has solutions of the form

$$\mathscr{R}(\rho) \sim A \exp\left(\frac{-\rho}{2}\right) + B \exp\left(\frac{\rho}{2}\right)$$

Since we require the solutions to remain bounded as $\rho \to \infty$ we set $B = 0$, and we therefore look for a solution of the form

$$\mathscr{R}(\rho) = e^{-\rho/2}K(\rho) \qquad \text{(VII.2)}$$

where $K(\rho)$ must vary less rapidly than $e^{\rho/2}$ if $\mathscr{R}(\rho)$ is to remain bounded at infinity. Substitution of equation (VII.2) in equation (VII.1) yields

$$\frac{d^2K}{d\rho^2} + \left(\frac{2}{\rho} - 1\right)\frac{dK}{d\rho} + \left\{\frac{\lambda - 1}{\rho} - \frac{l(l+1)}{\rho^2}\right\}K = 0 \qquad \text{(VII.3)}$$

The solution to this equation must remain valid right down to $\rho = 0$, where there is a singularity. We therefore write

$$K(\rho) = \rho^s \sum_{j=0}^{\infty} a_j \rho^j$$

$$= \rho^s L(\rho)$$

(VII.4)

where s is a positive number. Determining K'' and K' and substituting in equation (VII.3), comparison of coefficients of ρ^s yields the following *indicial equation*

$$\{s(s-1) + 2s - l(l+1)\}a_0 = 0$$

If $a_0 \neq 0$, then

(VII.5)

$$s = l \quad \text{or} \quad -(l+1)$$

We choose $s = l$ since the other solution introduces a further singularity in the region of the origin. We now have

$$K(\rho) = \rho^l L(\rho)$$

(VII.6)

and it remains to determine the coefficients in the infinite series $L(\rho)$, equation (VII.4).

Substituting equation (VII.6) in equation (VII.3) we obtain the following differential equation for $L(\rho)$:

$$\rho \frac{d^2 L}{d\rho^2} + \{2(l+1) - \rho\} \frac{dL}{d\rho} + (\lambda - l - 1)L = 0$$

where

(VII.7)

$$L(\rho) = \sum_{j=0}^{\infty} a_j \rho^j$$

Comparison of coefficients yields the *recurrence relation* between adjacent coefficients:

$$a_{j+1} = \frac{(l+j+1) - \lambda}{(j+1)(j+2+2l)} a_j$$

(VII.8)

Consider the ratio of two successive coefficients. We find

$$\frac{a_{j+1}}{a_j} \to \frac{1}{j} \quad \text{as} \quad j \to \infty$$

For large values of j the series $L(\rho)$ behaves like $\sum_j (\rho^j/j!)$, which is the series expansion of e^ρ. $\mathcal{R}(\rho)$ evidently behaves like $\rho^l e^{\rho/2}$ which diverges at infinity. Clearly, for acceptable (bounded) solutions we must require the infinite series to terminate in the form of a polynomial. Thus if in equation (VII.8) we set

$$\lambda = (l+j+1)$$

(VII.9)

the series will terminate at the jth term, all subsequent coefficients $a_{j+1}, a_{j+2}, \ldots$, being zero. Since both l and j are integral positive numbers, we may set

$$n \equiv j + l + 1 \qquad j = 0, 1, 2, \ldots$$
$$l = 0, 1, 2, \ldots$$
(VII.10)

where $n \geqslant 1$. This therefore restricts λ to adopt integral values, and effectively quantizes the total energy of the system. From Section 4.7 we may see, for example, that

$$E = -\frac{Z^2 e^4 m_e}{2\hbar^2 (j + l + 1)^2} = -\frac{Z^2 e^4 m_e}{2\hbar^2 n^2} \quad n = 1, 2, \ldots \qquad \text{(VII.11)}$$

which is an explicit statement of the energy quantization.

If, by $L_k(\rho)$ we mean that Laguerre polynomial which terminates at the kth term, then we may easily show:

$$L_0(\rho) = 1$$
$$L_1(\rho) = 1 - \rho$$
$$L_2(\rho) = 2 - 4\rho + \rho^2$$
$$L_3(\rho) = 6 - 18\rho + 9\rho^2 - \rho^3$$

These functions are shown in Figure VII.1. The generating function is

$$L_k(\rho) = e^\rho \frac{d^k}{d\rho^k} (e^{-\rho} \rho^k) \qquad \text{(VII.12)}$$

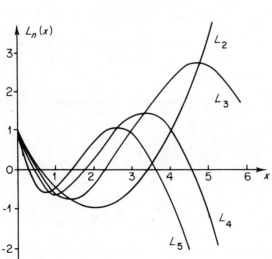

Figure VII.1

which may be directly verified by substitution in equation (VII.7). We may generate the *associated Laguerre polynomials* as follows:

$$L_k^r(\rho) = \frac{d^r}{d\rho^r}(L_k(\rho)) \qquad \text{(VII.13)}$$

where $L_k^r(\rho)$, the rth derivative of $L_k(\rho)$, is of degree $k - r$ and order r. If we take $L_k^{r''}$, $L_k^{r'}$ and substitute in equation (VII.7) we obtain

$$\rho L_k^{r''} + (r + 1 - \rho)L_k^{r'} + (k - r)L_k^r = 0 \qquad \text{(VII.14)}$$

which becomes identical to equation (VII.7)

$$\rho L'' + (2(l + 1) - \rho)L' + (n - l - 1)L = 0$$

provided we set

$$2(l + 1) = r + 1$$
$$n - l - 1 = k - r \qquad \text{(VII.15)}$$

Solving for k and r we obtain

$$k = n + l$$
$$r = 2l + 1 \qquad \text{(VII.16)}$$

Hence the final radial solution is

$$\mathscr{R}_{nl}(\rho) = e^{-\rho/2} \rho^l L_{n+1}^{2l+1}(\rho) \qquad \text{(VII.17)}$$

Since $\mathscr{R}_{nl}(r)$ contains a polynomial of order $k = n - l - 1$, the curve representing it will have k roots and will consequently cross the r-axis k times in the region $0 \leqslant r \leqslant \infty$. In Figure 4.7.1 we show the qualitative form of the radial probability distribution $4\pi r^2 \mathscr{R}_{nl}^2(r)$.

The component ρ^l looks after the singularity at the origin, whilst $e^{-\rho/2}$ ensures that the asymptotic form of the radial solution remains bounded, and more or less located around the origin.

If we now consider the *positive* energy solutions corresponding to the unbound states, the separated radial equation becomes

$$\frac{d^2\mathscr{R}}{d\rho^2} + \frac{2}{\rho}\frac{d\mathscr{R}}{d\rho} - \left\{ -\frac{1}{4} - \frac{\lambda}{\rho} + \frac{l(l + 1)}{\rho^2} \right\}\mathscr{R} = 0 \qquad \text{(VII.18)}$$

Again, the asymptotic form is

$$\frac{d^2\mathscr{R}}{d\rho^2} + \frac{1}{4}\mathscr{R} \sim 0 \qquad \text{(VII.19)}$$

which this time has oscillatory and unattenuated solutions

$$\mathscr{R}(\rho) \sim A e^{\pm i\rho/2} \qquad \text{(VII.20)}$$

Equation (VII.18) has a pole at the origin, and we therefore assume a solution of the form

$$\mathcal{R}(\rho) = e^{\pm i\rho/2} \rho^s \mathcal{L}(\rho) \qquad \text{(VII.21)}$$

Establishing the indicial equation as before, we find $s = l + 1$, this ensuring the removal of the pole at the origin. I.e.

$$\mathcal{R}(\rho) = e^{\pm i\rho/2} \rho^{l+1} \mathcal{L}(\rho) \qquad \text{(VII.22)}$$

If we attempt a series solution

$$\sum_j a_j \rho^j \text{ of } \mathcal{L}(\rho)$$

as we did in the bound case, we obtain the following recursion

$$a_{j+1} = \frac{\mp i(j + l + 1) - \lambda}{(j + 1)(j + 2l + 2)} a_j \qquad \text{(VII.23)}$$

where $i = \sqrt{-1}$. For large values of j the series behaves like $e^{\mp i\rho}$, which may be normalized at infinity, and therefore there is no need to terminate the series as a polynomial. Consequently there is no restriction on λ, and hence the total energy, and so the unbound states are unquantized. The wavefunctions form a discrete spectrum of states in the bound case, and a continuum of states in the unbound case.

Normalization

The radial bound solutions $\mathcal{R}(\rho)$ may be normalized, and the normalization constant is given as

$$N_r(n, l) = \left[\int_0^\infty R_{nl} R_{nl} r^2 \, dr \right]^{-\frac{1}{2}}$$

$$= \sqrt{\frac{4(n - l - 1)!}{a^3 n^4 [(n + l)!]^3}} \qquad \text{(VII.24)}$$

where a is the Bohr radius of the first quantized orbit.

Bibliography

The texts listed below provide a more detailed and advanced treatment of certain physical and mathematical aspects of the material developed in this book. This bibliography is by no means exhaustive and is meant simply to direct the reader to those texts and references for which this book provides a working foundation. Nevertheless these books, in the author's experience, represent readable extensions to the present work and are commended to those requiring a more extensive treatment.

General Advanced Texts

The following texts provide a highly detailed and exhaustive review of the physical application of advanced mathematical techniques.

1. Morse, P. M., and Feshbach, H., *Methods of Theoretical Physics* (2 Vols.), McGraw–Hill, New York (1953) provides a wide-ranging treatment of scalar and vector field problems and the development of the Green's functions of several problems of classical physics.
2. Killingbeck, J., and Cole, G. H. A., *Mathematical Techniques and Physical Applications*, Academic Press, New York (1971). An extensive review of a broad range of mathematical techniques applied to a large number of advanced physical problems.
3. Joos, G., *Theoretical Physics*, third revised edition, Blackie, Glasgow and London (1958). Now a rather dated text, but nevertheless provides a very sound theoretical introduction to a number of physical problems.
4. Kompanyets, A. C., *Theoretical Physics*, Foreign Languages Publishing House, Moscow (1961).
5. Landau, L. D., and Lifshits, E. M., *A Course of Theoretical Physics* (9 Vols.), Pergamon Press, Oxford (1959–1969). An exhaustive *tour de force* with volumes devoted to classical and quantum mechanics, statistical physics, fluid mechanics, elasticity, electrodynamics and, in particular, the classical theory of fields.
6. Courant, R., and Hilbert, D., *Methods of Mathematical Physics* (2 vols.), Wiley, New York (Vol. 1, 1953, Vol. 2, 1962).
7. Mathews, J., and Walker, R. L., *Mathematical Methods of Physics*, 2nd edition, Benjamin, New York (1970). A useful and readable book with a mathematical rather than physical emphasis, but nevertheless, appropriate to physics.
8. Jeffreys, H., and Jeffreys, B. S., *Methods of Mathematical Physics*, 3rd edition, Cambridge University Press, London and New York (1956).

Chapter 1

There are a large number of texts which provide a sound basic exposition of scalar and vector field theory. Of the general references, Morse and Feshbach (1), Killingbeck and Cole (2) and Joos (3) are recommended. In addition

9. McQuistan, R. B., *Scalar and Vector Fields*, Wiley, New York (1965), provides a detailed physical interpretation of the vector operators and the basic features of scalar and vector fields.

10. Hague, B., *An Introduction to Vector Analysis*, Methuen, London (1961). A short introductory monograph containing the basic results of vector analysis. Another monograph in the same category is
11. Rutherford, D. E., *Vector Methods*, Oliver and Boyd, London (1939)
12. Queen, N. M., *Vector Analysis*, McGraw–Hill, London and New York (1967). Good treatment of cartesian vectors and tensors.
    Good discussions of orthogonality are given by Mathews and Walker (7), Killingbeck and Cole (2), Morse and Feshbach (1). An elementary discussion of orthogonality is given by
13. Kauzmann, W., *Quantum Chemistry*, Academic Press, New York (1957) and by
14. Coulson, C. A., *Valence*, Oxford University Press (1961).
    A detailed discussion of Sturm–Liouville theory and completeness is given by
15. Margenau, H., and Murphy, G. M., *Mathematics of Physics and Chemistry*, Van Nostrand, Princeton (1967). See also Courant and Hilbert (6).
    A good discussion of uniqueness is that of
16. Kellogg, O. D., *Foundations of Potential Theory*, Ungar, New York (1949).

Chapter 2

Separation of variables is treated at some length in Morse and Feshbach (1) who also extend the separation to the general coordinate frame utilizing Stäckel determinant.

Hybridization is discussed in Kauzmann (13), Coulson (14) and Killingbeck and Cole (2) for a variety of classical and quantum mechanical systems. See also

17. Fong, P., *Elementary Quantum Mechanics*, Addison–Wesley, London (1962).
    A general reference giving a discussion of a number of classical vibrating systems is
18. Morse, P. M., *Vibration and Sound*, 2nd edition, McGraw–Hill, New York (1948) and
19. Coulson, C. A., *Waves*, 6th edition, Wiley, New York (1955).
20. Brillouin, L., *Wave Propagation in Periodic Structures*, McGraw–Hill, New York (1960).
    Propagation of lattice waves in disordered structures is discussed by
21. Chernov, L. A., *Wave Propagation in a Random Medium*, McGraw–Hill, New York (1960).
    Fong (16) provides a good discussion of the solution of the one-dimensional Schrödinger equation; a more concise treatment is given by
22. Matthews, P. T., *Introduction to Quantum Mechanics*, McGraw–Hill, New York (1963).
    For the discussion of a variety of simple quantum mechanical problems see
23. Evans, R., *The Atomic Nucleus*, McGraw–Hill, New York (1955).
    Applications of the Mathieu equation are given by Morse and Feshbach (1), Mathews and Walker (7) and Brillouin (20). See also
24. McLachlan, N. W., *Theory and Application of Mathieu Functions*, Dover, New York (1947).

Chapter 3

25. Watson, G. N., *A Treatise on the Theory of the Bessel Functions*, 2nd edition, Cambridge University Press, New York and London (1948), and
26. Abramowitz, M., and Segun, I. A., *Handbook of Mathematical Functions*, Dover, New York (1965) provide full discussions of the mathematical properties and tabulations of the Bessel functions. Another standard tabulation is that of
27. Jahnke, E., and Emde, F., *Tables of Functions*, 3rd edition, Dover (1943).
    For another discussion of correlation amongst square and circular normal modes see Kauzmann (13).

Chapter 4

Various aspects of potential theory, in particular the application of spherical harmonic functions are discussed by Morse and Feshbach (1), Mathews and Walker (7).

28. Webster, A. G., *Partial Differential Equations of Mathematical Physics*, 2nd edition, Dover, New York (1933).
29. Goertzel, G. H., and Trelli, N., *Some Mathematical Methods of Physics*, McGraw–Hill, New York (1960).
30. Jeans, J. H., *Mathematical Theory of Electricity and Magnetism*, 5th edition, Cambridge University Press, London and New York (1925).
31. Sternberg, W. J., and Smith, T. L., *The Theory of Potential and Spherical Harmonics*, University of Toronto Press, Toronto (1952).

Application of the Laplace equation to problems of hydrodynamics is discussed in

32. Lamb, H., *Hydrodynamics*, 6th edition, Dover, New York (1932).

For a somewhat simpler and more restricted discussion see

33. Bleaney, B., and Bleaney, B. I., *Electricity and Magnetism*, Oxford University Press (1965).

The general properties of the spherical Bessel functions are developed in Watson (25) and are tabulated in Abramowitz and Segun (26) and Jahnke and Emde (27).

Scattering is treated in some detail by Matthews (22) and Evans (23). Landau and Lifshits (5) in their book on Statistical Physics give some discussion of phase shift analysis as does Matthews (22) and

34. Faber, T. E., *Liquid Metals*, Cambridge University Press, London and New York (1972).

A very full discussion of atomic spectra and their modification in the presence of electric and magnetic fields is given by

35. White, H. E., *Atomic Spectra*, McGraw–Hill, New York (1934).

In addition to Faber's monograph (34),

36. Egelstaff, P. A., *Introduction to Liquid State Physics*, Academic Press, New York (1967), and
37. March, N. H., *Liquid Metals*, Pergamon Press, Oxford (1968) can be recommended
for a discussion of the liquid metal pseudopotential. Oscillations of a planet's atmosphere taking into account thermal and gravitational effects may be discussed in terms of the Laplace equation. See, for example

38. Wilkes, M. W., *Oscillations of the Earth's Atmosphere*, Cambridge University Press, London and New York (1949).

Morse and Feshbach (1) and the references contained therein give a more extensive discussion of the application of spheroidal coordinates to physical problems.

Chapter 5

There are a number of books dealing with the variational methods. That of

39. Lanczos, C., *The Variational Principles of Mechanics*, 3rd edition, University of Toronto Press, Toronto (1966) is probably the best for mechanical applications of the variational principle.
40. Mikhlin, S. G., *Variational Methods in Mathematical Physics*, Pergamon Press, Oxford (1964) gives a general discussion of the variational principle applied to a number of physical problems, based on the energy method. Rather more advanced is
41. Yourgrau, W., and Mandelstam, S., *Variational Principles in Dynamic and Quantum Theory*, 2nd edition, Pitman, New York (1960).

Further discussions, and references to numerous other texts dealing with the calculus of variations are to be found in the general references: Morse and Feshbach (1), Killingbeck and Cole (2), Mathews and Walker (7), Kauzmann (13), Coulson (14).

Index